P9-AFK-899

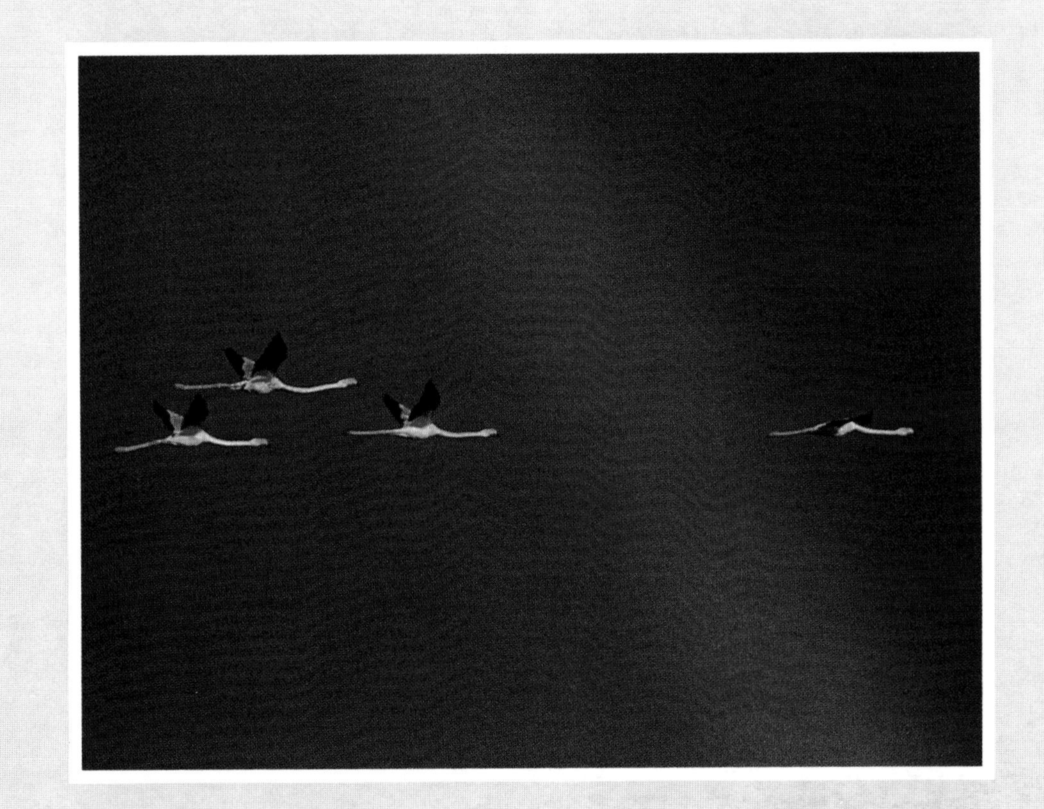

COLORS
IN THE WILD

NATIONAL WILDLIFE FEDERATION

Library of Congress CIP Data: page 155.

COLORS. They can delight the eye, create a mood, set human emotions soaring in any of nature's seasons. In winter, the cardinal looks spectacular, more intensely red than usual against the bleak sky. Come spring, a tiger swallowtail floats lazily amid the purple blossoms of cress. Then summer's long days begin to turn the greens deeper, the reds and oranges more vivid. Fall's cooler mornings are warmed by the blazing splendor of red maples and sweet gum trees; brown branches are finally laid bare against the late autumn sky.

In the world of nature, colors play a practical role. The most important color, it could be argued, is the green pigment, chlorophyll. Without chlorophyll, plants could not capture sunlight and turn it into food. Without plants, animals—including man—would die. But color is a tool for survival in many other ways, too.

Colors advertise: sex, food. They conceal: prey from predators, and vice versa. And they warn: bad taste, danger, no trespassing. In short, color helps animals exist in particular environments and communicate with other living things.

The glossy sheen of buttercup petals alerts insects to the presence of nectar. As an insect feeds, its wings are dusted with the flower's pollen. Then it flits to another flower, carrying the pollen along. Plants that need to attract birds or insects for pollination thereby get pollinated effectively, thanks in part to color's role in the process.

Bright colors indicate to insects that nectar is at hand. Scientist Karl von Frisch showed that color was the attraction and that bees could even discern different shades. He trained honeybees to feed at a bowl of sugar-water placed on blue paper. When he gave the bees a choice between gray and blue paper, they still flew to the blue paper—even if no food was on it.

Colors also guide an insect over the flower at close range. Colored lines on the petals—often ultraviolet and thus invisible to humans—have a "bull's eye" effect. They help the insect zero in on the nectar, making sure the insect touches the blossom's anthers and stigmas to complete pollination and perpetuate the species.

Among animals, social communication, aided greatly by color, keeps many species going—breeding and feeding and protecting territory. During the breeding season, male frigate birds inflate an enormous red throat pouch to attract passing females. In an intricate, high-stepping mating display, male and female blue-footed boobies strut toward each other, waggling their bright blue feet. As spring approaches, the colors of male birds, such as cardinals and goldfinches, intensify in brightness and allure to female birds.

Color can help an animal protect its territory. When a male robin catches sight of another male robin entering its territory, it turns its red breast toward the rival and flies upward to display the greatest possible area of red. The red signal warns intruders to keep their distance. If one ignores the signal, the resident robin responds angrily. One scientist showed that a robin will attack a mere patch of red feathers, even without a bird attached, if such an object comes into its territory during the mating season.

These color signals that aid communication are known as "releasers" because they elicit an instinctive response. In one experiment, a scientist showed that the red belly of a male three-spined stickleback, a small fish, provoked aggressive responses in other males. Even crude models of a stickleback, if painted red, provoked more attacks than a realistic stickleback model without a red belly.

In the day-to-day business of life, consider also the role of other bright colors. When a hungry baby robin stretches open its scarlet mouth, parent birds instinctively stuff food into the gap. Almost any brightly-colored, triangular-shaped cavity triggers this response, researchers have found, including cardboard models. Without the triangular patch of color to advertise its need for food, even the robin's plaintive cries will not prompt its parents to feed it. Many other birds respond in the same way to shapes and to patches of color. Adult zebra finches recognize juveniles by their black bills and, in one experiment, did not feed them when their bills were painted red, even when the young ones cried pitifully.

While color helps preserve a species by bringing mates together and keeping young ones fed, it is just as intriguing when used *against* other animals, either when animals try to escape the notice of predators, or when predators wait to ambush prey. Wearing drab gray or brown, many animals better avoid predators by concealing themselves against rocks, earth, or vegetation. The mottled brown and buff of a woodcock's plumage, for instance, blend with surrounding vegetation and conceal it from a fox. While a tiger waits for prey to come along in the jungle, its stripes blend into the dappled patterns of light, hiding the lurking animal.

Odd as it may seem, bright, conspicuous patterns can also "hide" an animal. The bold markings distract the observer's eye, making the viewer overlook the outline of the animal itself. A broad black stripe encircling the body of a red fish, for instance, makes it look more like three separate blobs, none of which quite resembles a fish.

Abbott Thayer, an artist and avid naturalist in the early 1900s, did much to show just how valuable color is in helping animals escape detection. He used his paintings, such as those of a peacock in the forest and a copperhead resting on dead leaves, to show that protective coloration often made animals nearly invisible.

Scientists agreed that the flecked coloring of woodland birds resembled the dappled effect of sunlight through bits of branches. But past that, fierce arguments raged.

An intense, determined man, Thayer often tried to prove his points by collecting a crowd for an outdoor demonstration. In Washington, he tried to convince a group that the white patches on pronghorns conceal them when the animals are viewed from a certain angle. When Thayer told the men they had to lie on the muddy earth and peer upward to appreciate this angle, the demonstration ground to a halt.

Thayer, however, never lost faith in his theories, despite heavy criticism. After he argued that flamingoes were a rosy color in order to conceal them at sunrise and sunset, Theodore Roosevelt took him to task. The times when flamingoes might be shielded by the light of dawn or dusk were too few to be worth considering, Roosevelt reasoned. Tongue in cheek, the former president suggested that even the raven's coloration could be called concealing if "it is put into a coal scuttle."

Thayer's reputation today rests on his more substantial theory of countershading, which says that animals tend to be colored darkest on those parts of their bodies that tend to be most exposed to the sunlight, and lightest on the part in the shadow. When light hits the dark parts and shadow hits the light parts, the tones cancel each other out, obscuring the animal. To see the theory at work, look at the gradations of color in a deer, for instance.

During World War I, Thayer tried to persuade the Allies to adopt some of his theories in painting ships and designing uniforms. Though he got a lukewarm response from military leaders, his ideas set others to thinking and experimenting. During World War II, when aerial warfare became more common, his theories of camouflage and countershading found practical application at last.

Animals also sometimes adapt their coloration to match changing environmental conditions. In England before 1850, a few peppered moths were black, but most were a pale gray with flecks of black. The gray color blended well with the lichen-covered tree trunks on which the moths habitually rested. Then came the factories with their smoke. The pollution killed off the lichens and blackened the tree trunks: the gray peppered moth's coloration now threatened to be its undoing. The black form of the moth had a better chance of surviving against the black tree trunks, and by the end of the century, the black moth became common. Color was the key to its population explosion.

To prove it, H.B.D. Kettlewell released equal numbers of pale and black moths in an area with light-colored tree trunks. Birds ate six times as many black moths as pale ones. But when he did the same thing in an industrial area, black moths survived pale ones by 2 to 1, because birds could more easily spot the pale moths on the trees' darkened bark. Today, because of cleaner industry, England's trees are changing back to a paler color. Likewise, a paler form of the moth now holds the edge.

Color changes can evolve over the years—or they can happen before our eyes. The chameleon's quick-change artistry is well-known, yet octopuses and cuttlefish can switch color even faster, in about two-thirds of a second. The American horned lizard, when moved from pale desert soil to reddish-brown earth, changes to the color of the new soil in a few hours.

Seasonal color changes are almost as striking. The snowshoe hare that traveled about in summer's dry grass now molts into a white coat and hops with near impunity atop the winter snow.

The same species can even take different color forms in different environments. Yellow-wattled plovers lay reddish eggs in an area of brick-red laterite on the Indian coast, but they lay predominantly brown eggs in the dark soil of the surrounding country. In New Mexico, a dark version of the pocket mouse lives on a black lava flow in the Alamogordo region, while tawny forms of the mouse live in the sere desert nearby.

Such color camouflage allows an animal to blend with its background. Disguise, on the other hand, uses the animal's shape *and* color to make it look like some inanimate part of its environment, such as a leaf or a twig. Disguised animals rely on not being recognized for what they really are, and that makes them less likely to be detected as they lie in wait for prey, or less likely to become a predator's next meal.

A predator often can't tell a tent caterpillar from tree bark. A mottled brown treehopper in Mexico looks for all the world like a gnarled tree limb. In tropical waters, the stone fish resembles a chunk of rock—except when it leaps forward to kill unwary prey. In tropical rain forests, a leaf mantid's veined wings expand and contract into leaf-like structures, and stick caterpillars resemble small twigs growing on the trees on which they feed.

The most bizarre animal disguises may be the moths and caterpillars that bear a disgusting resemblance to blobs of bird-droppings. One early explorer put it this way: "I was stretching across to collect a beetle and nearly touched what I took to be the excrement of a crow. Then to my astonishment I saw it was a caterpillar half-hanging, half-lying on a leaf. Another thing that struck me was the skill with which the coloring rendered the varying surfaces, the dried portion at the top, then the main portion, moist, viscid, soft, and the glistening globule at the end. A skilled artist, working with all the materials at his command, could not have done it better."

Disguised creatures may take one tack to avoid detection, but other animals take quite another: warning coloration. A frog sporting a flaming red vest and blue legs hops along the forest floor, its bold colors warning predators of its nauseating taste. The bright colors of the blister beetle remind predators that it secretes a caustic liquid which blisters the skin of any animal that rubs against it. A bird's memory of the terrible taste of a noxious butterfly protects all others of that butterfly species from at least *that* hungry bird.

An animal with warning colors often has a toxic property strong enough to repel, but not strong enough to kill. If the predator dies on the first encounter, it has no chance to learn by experience. Only after predators have attacked an animal with warning colors and suffered the consequences will they learn to associate bright color patterns with unpleasant experiences. Combinations of black, red, orange, and yellow are common; metallic blues or greens are also effective.

We know that predators recognize offensive-tasting prey by the prey's coloration rather than by some other clue. The proof is provided through "mimics," harmless animals whose colorations mimic those of harmful ones. Most familiar is the monarch butterfly, whose mimic, the viceroy, looks almost identical except in size. Birds dislike the monarch because it tastes bad. Its bold color pattern warns birds not to attack it. The viceroy doesn't taste bad, but birds still tend to leave it alone because its colors remind them of the bad-tasting monarch.

The same is true for some species of clearwing moths, which display wasp-like bands of yellow or orange-red around the body. They even buzz ominously as they fly. The clearwing moth can't sting, but the wasp-like get-up is enough to scare off most birds.

Some snakes sport the same colors as poisonous coral snakes but in slightly different combinations. Yet predators (and hiking humans) give a wide berth to almost any slender snake with rings of black, red, and yellow.

To warn, to conceal, to attract, to protect: color is essential for all these purposes. Yet what, one may ask, creates color in the first place, and what makes any animal or plant the color it is?

Light waves strike the earth continuously. The shortest of the waves people can see produce blue colors; the longest, red, with the rest of the spectrum in between. Most creatures' vision has adapted to better see certain color ranges, which helps them survive in their particular environments. Bees have better color vision in the blue end of the spectrum, but they cannot detect red as we humans know it. They can, however, see ultraviolet waves, which means they see many colors unimaginable to us. On many flowers, the ultraviolet tones act like landing lights for a bee, allowing it to better home in on precious nectar. Freshwater fish have color vision shifted to the red end of the spectrum; saltwater fish have better color vision at the blue end. The differences evolved because of the different colored water in which each needs to see.

Apes, monkeys, most birds, lizards, bony fish, and insects all have good color vision to record social signals. Most birds see blue colors poorly, but see red ones well. Accordingly, many plants that depend on birds for pollination have evolved red flowers or seeds wrapped in red fruit to entice the birds.

No plant or animal chooses what color it wants to see best or what color it wants to be. Individuals which blend best with their surroundings have often had the advantage. If more animals of one design survive than those of another, that design tends to mark the species. In some cases, a particular red or orange petal color has proven more effective in attracting certain birds to pollinate, so that petal color has come to mark a certain plant species. Flower colors which attracted the right animals most successfully are the ones that have survived. Over the years, animals and plants have evolved color characteristics that better fit their different environments.

Some birds, most often sea birds that feed from the water's surface, have in their eyes colored oil droplets that filter out the red portion of the spectrum. That makes it possible for gulls and terns, for example, to see better in their foggy and cloudy seashore environments. The filter makes the color white stand out clearly—so clearly, in fact, that it may also help these birds more easily spot plankton, and thereby find fish to eat.

Myriad colors abound, but they all derive from two sources—structure and pigment. Structural color is created by the physical nature of a surface. When light hits a finely-textured surface or a thin, transparent layer, it bends and scatters into waves of red, orange, yellow, and so forth. When light hits a butterfly's wing, the surface of the wing reflects all the colors contained in visible light, but it reflects *more* of certain colors, and these are the colors we see. Human eyes record a brilliant yellow wing, for instance, because more yellow waves are reflected than anything else.

The white of lily flowers also is structural. Air spaces within the plant's tissue do not absorb one wavelength more than any other, so the petals appear white. Snow, a mass of crystals and air spaces, works the same way.

A structural color known as Tyndall blue causes many of the brilliant blues in nature. John Tyndall, a 19th century English physicist, found that fine particles of dust in the atmosphere scatter the different wavelengths of light that comprise sunlight. The shorter wavelengths at the blue end of the spectrum, which are scattered most, are viewed against the black background of space. That is why—all poetry, painting, and sentiment aside—humans say the sky looks blue.

Tyndall blue shows up in animals the same way. The kingfisher, for instance, has in its feathers tiny air bubbles that scatter and reflect blue wavelengths, which are then visible against a background of black melanin, a pigment.

The second type of color, pigmentary, is created in

animal and plant cells by chemicals that selectively absorb or reflect certain colors. Pigments produce some greens and many blacks, browns, reds, and yellows.

The color of any animal may be pigmentary, structural, or a combination. A frog that appears green to us has a background pigment of yellow, along with specialized cells that scatter blue light. The blue light mixed with the yellow background makes the frog look green.

The color of a pigment is due to its absorbence of all other wavelengths of light. Chlorophyll absorbs red and blue, for instance, and appears to us as green. Likewise, pigments are what make a flamingo pink. The pink comes from algae the flamingo ingests from the alkaline waters of the lakes in which it lives. The algae contains a reddish carotenoid pigment. Put a lake-living flamingo in a zoo, and the pink color will disappear—unless the zoo knows (which it does) to adjust the flamingo's diet accordingly during its molt. The rich orange color of crab eggs is also produced by a carotenoid pigment, the same one that makes salmon pink and egg yolks yellow.

Yellow, orange, and red carotenoids, in fact, help humans see color in the first place. Animals depend on them for vitamin A, essential to good eyesight. Mammals cannot manufacture carotenoids, so they rely on carotenoids made by plants. They get them firsthand by eating the plants (that's why parents exhort kids to eat their carrots) or secondhand by eating herbivorous animals.

Pigments may be present in skin or fur from birth, concentrated in a layer of granules or droplets beneath the surface. Or they exist in hair, feathers, or scales. From the freckles of humans to the downy yellow plumage of chicks, pigments are responsible.

It is pigmentary color cells that allow chameleons, octupuses, and a few other animals to put on their fast-changing color shows. These pigments are contained in special cells that can be contracted or expanded to either reveal or hide the pigment. In cuttlefish, pigments and cells bunch up, then scatter, so that rich reds and browns scuttle across their skin in a matter of seconds.

If the pigment granules are a dark melanin, they absorb all the light when the granules are spread out, and the skin appears dark. When the cells contract, lighter pigments prevail—and the animal changes color.

Even the spectacular show of autumn leaves is caused by several pigments. When the green pigment (chlorophyll) has no more work to do, it breaks down, unmasking the yellow, orange, and red pigments. The pigments were in the leaf all along, hidden by the higher concentration of chlorophyll. As the green pigment disappears, a process usually triggered by cool weather, the leaves put on their splendid change of color.

No matter how much we know about how color is created, scientific explanations cannot rob color of its wonder. The beauty of color appeals to us all. The insightful Helen Keller, blind and deaf since shortly after birth, was once asked how she perceived the idea of color. "I habitually think of things as colored and resonant," she answered. "I understand how scarlet can differ from crimson because I know that the smell of an orange is not the smell of grapefruit. I can also conceive that colors have shades and guess what the shades are."

All of us might take a closer look at nature, where form, texture and color combine to produce incomparable beauty, a beauty we celebrate in *Colors in the Wild.*

COLORS

AN AUTUMN RED MAPLE LEAF IN A WASHINGTON STATE FOREST PROVIDES A CRIMSON PLATFORM FOR A TINY TREE FROG *(LEFT)*. ANOTHER FOREST DWELLER, CLADONIA LICHEN *(RIGHT)*, GROWS MOSTLY ON LOGS AND TREE TRUNKS. ITS PUFFY RED FRUITING CAPS HAVE EARNED IT THE NICKNAME BRITISH SOLDIER.

PREVIOUS PAGE: A RED MAPLE LEAF DRIFTING IN A WOODLAND STREAM SIGNALS WINTER'S APPROACH.

THE BLACK SWAN'S EXOTIC BEAUTY *(LEFT)* IS PROBABLY WHAT PROMPTED WINSTON CHURCHILL AND THE EMPRESS JOSEPHINE TO IMPORT THE BIRDS FOR PRIVATE PONDS AND PARKS. ONLY IN AUSTRALIA AND NEW ZEALAND, HOWEVER, ARE THERE WILD FLOCKS.

A "BLOOD STAR" STARFISH *(RIGHT)* VIVIDLY LIVES UP TO ITS NAME, ESPECIALLY WHEN SEEN AGAINST DARK, SUN-DRIED KELP.

THE SEA IS RICH IN BRIGHTLY-HUED LIVING THINGS. SEA FAN CORAL ATTRACTS A PIPEFISH, WHICH SEARCHES FOR FOOD IN THE MAZE OF CARMINE BRANCHES *(FAR LEFT)*. MEANWHILE, A SCARLET AND WHITE SHRIMP—APTLY CALLED A WHITE-BOOTED SHRIMP—PICKS ITS WAY ALONG THE OCEAN FLOOR *(LEFT)*.

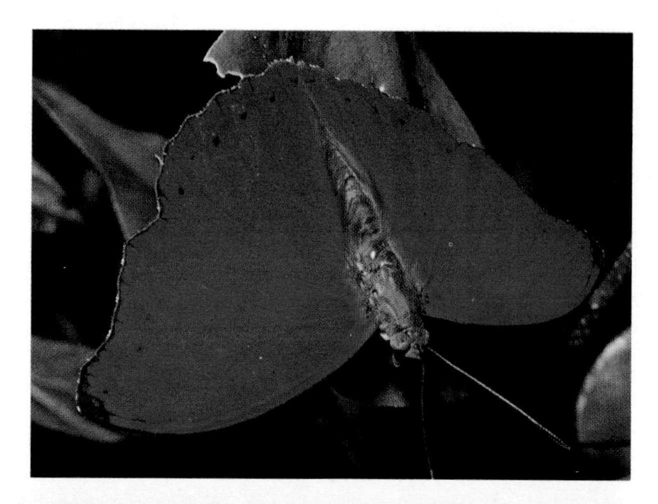

VIVID PLANTS AND ANIMALS BRIGHTEN THE SHADED WORLD OF TROPICAL RAIN FORESTS. FIERY LEAVES OF A FLAME BROMELIAD *(RIGHT)* FORM A CUPLIKE CENTER WHERE RAINWATER COL-LECTS, AND A CYMOTHOE BUTTERFLY'S RUBY WINGS *(TOP)* FLASH WHEN THEY FLUTTER. THE RED-EYED LEAF FROG'S INTENSELY-TINTED EYES *(ABOVE)* MAY HELP TO STARTLE AWAY POTENTIAL PREDATORS.

A SOCKEYE SALMON, SEEN HERE IN ITS ROSEATE SPAWNING-SEASON COLOR PHASE, MAKES A CHOICE MEAL FOR AN ALASKAN BROWN BEAR *(LEFT)*.

IN KENYA, A FLOCK OF LESSER FLAMINGOS PRANCES IN A FLORID COURTSHIP PARADE IN THE SHALLOWS OF LAKE NAKURU *(RIGHT)*.

TURNED SANGUINE FOR THE FALL SEASON, WINGED SUMAC DOMINATES A FOGGY, ABANDONED APPLE ORCHARD. IN SIMILAR FASHION, A MALE CARDINAL STANDS OUT SPLENDIDLY AGAINST THE GREYNESS OF AN OVERCAST WINTER DAY.

DESPITE ITS APPEARANCE, THE RED UAKARI MONKEY *(FAR LEFT)*, IS NEITHER SUNBURNED NOR FLUSHED WITH ANGER. THE VERMILION OF ITS FACE AND FOREHEAD IS A NATURAL, HEALTHY TRAIT IN THIS SPECIES. LOSS OF COLOR IS A SIGN OF ILLNESS.

RUDDY SKIN AROUND THE THROAT AND EYES OF AFRICA'S GROUND HORNBILL *(LEFT)* PROVIDES A DASHING TOUCH TO THE OTHERWISE DULL HUE OF THE TURKEY SIZED BIRD.

COLORS

PREVIOUS PAGE: AS IT FLITS FROM FLOWER TO FLOWER, A HONEYBEE LADEN WITH POLLEN IS AN UNWITTING AGENT IN THE POLLINATION PROCESS.

BLENDING WITH THE PETALS OF A BLACK-EYED SUSAN, A CRAB SPIDER *(ABOVE)* IS LESS OF A TARGET FOR PREDATORS. COLOR CAMOUFLAGE MAY ALSO HELP THIS WEBLESS SPIDER AMBUSH PREY.

A FOGGY FALL MORNING TAKES ON A WARM SUNNY FEELING WHEN VIEWED FROM THE EDGE OF A FIELD RESPLENDENT WITH GOLDENROD *(RIGHT)*.

YOUNG OLYMPIC MARMOTS DISCOVER A TREAT JUST OUTSIDE THEIR BURROW *(LEFT)*. WILDFLOWERS ARE QUITE OFTEN INCLUDED IN THEIR DIET OF HERBACEOUS PLANTS.

COME SPRING, WILLOW CATKINS *(RIGHT)* BURST FORTH LOOKING LIKE BUTTERY FUR BALLS, AND RECENTLY HATCHED GREAT EGRET CHICKS *(FAR RIGHT)* HUDDLE TOGETHER WHILE AWAITING DELIVERY OF A MEAL.

JACK-O-LANTERN MUSHROOMS *(LEFT)* ARE SO CALLED BECAUSE THEIR GILLS GIVE OFF A GREENISH GLOW IN THE DARK. THEY ARE POISONOUS TO HUMANS.

GROWING ON LOGS AND STUMPS OF CONIFEROUS TREES, YELLOW PHOLIOTA MUSHROOMS *(RIGHT)* ARE EDIBLE BUT HAVE LOOK ALIKES THAT ARE NOT.

THE DARK SHADES OF NIGHT ARE PIERCED BY THE LUMINOUS MOON ON AN AFRICAN PLAIN *(LEFT)* AND BY THE SULPHUR EYES OF A GREAT HORNED OWL *(FAR LEFT)* IN A WASHINGTON FOREST. IN YET ANOTHER HABITAT, THE DESERT OF ARIZONA, A PRICKLY PEAR CACTUS WEARS ITS SPLENDID BLOSSOM IN CELEBRATION OF SPRING *(ABOVE)*.

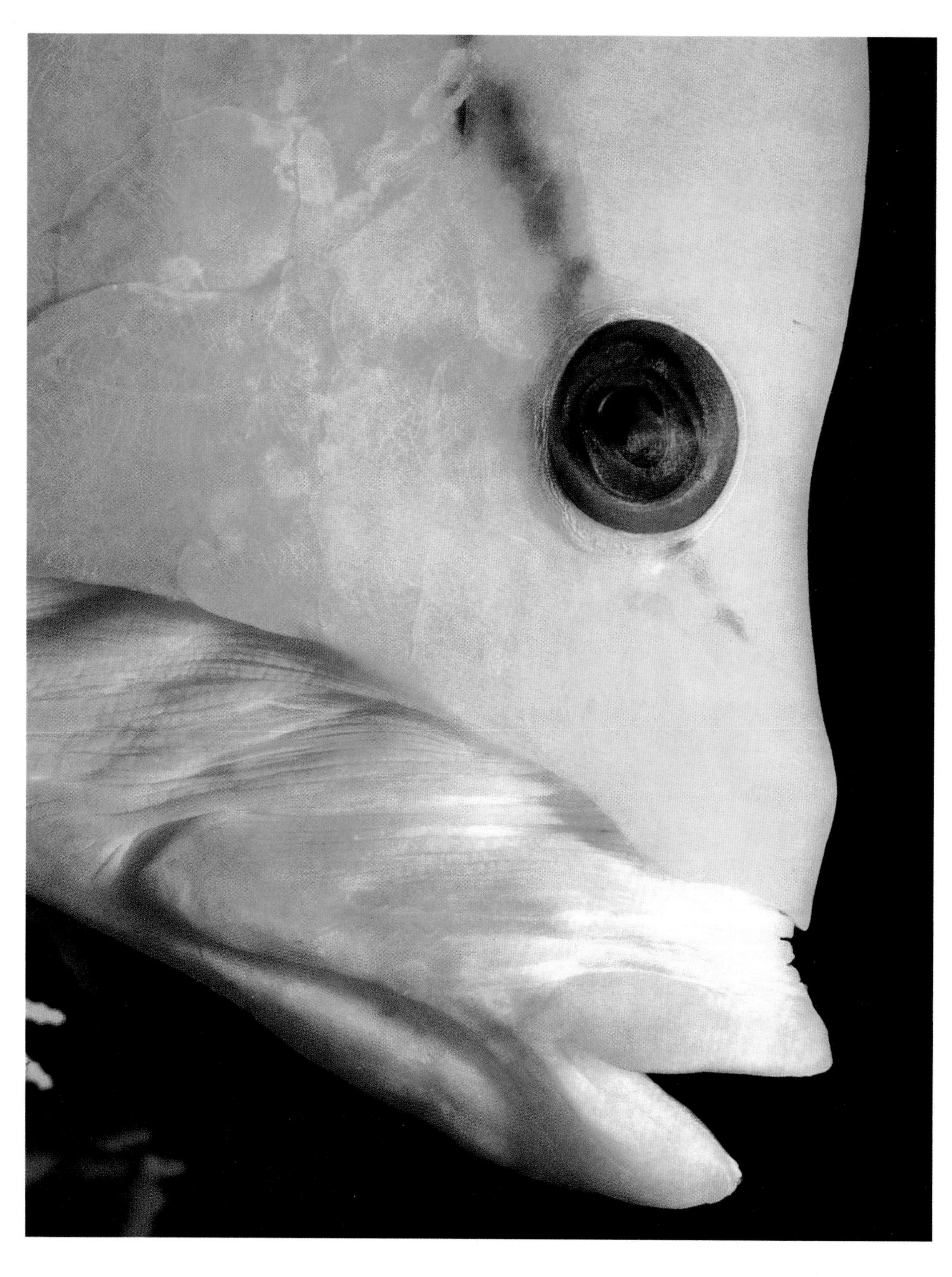

THE SLINGJAW WRASSE *(LEFT)* SNARES PREY BY SHOOTING ITS OPEN MOUTH FORWARD, CREATING A SUCTION THAT DRAWS IN MARINE ANIMALS THAT ARE SMALL ENOUGH AND CLOSE ENOUGH. NEARBY, THE BRANCHES OF CUP CORAL *(RIGHT)* WAVE IN THE RED SEA'S CURRENTS LIKE FLAXEN HAIR IN THE WIND.

THE CITRINE MARKINGS OF A SPOTTED SALAMANDER *(LEFT)* STAND OUT AGAINST ITS DARK BODY AS IT WALLOWS IN THE WELCOME MOISTURE OF A PUDDLE.

WHEN CONFRONTED, AUSTRALIA'S BEARDED DRAGON *(RIGHT)* INFLATES ITS THORNY THROAT SAC AND OPENS ITS MOUTH TO FLASH THE BRILLIANT LINING. SCIENTISTS BELIEVE IT STRIKES THIS POSE TO SCARE OFF AGGRESSORS.

SUGAR MAPLES IN MICHIGAN WEAR BOUFFANT BONNETS OF GOLDEN AUTUMN FOLIAGE. HASTENING TOWARD WINTER, SOME LEAVES HAVE ALREADY RELEASED THEIR GRIP ON THE BRANCHES. SCATTERED ON THE GROUND, THEY WILL DECAY AND REPLENISH THE SOIL FOR NEW GROWTH IN THE SPRING.

A CRESTED FLYCATCHER *(LEFT)* TRIUMPHANTLY CLUTCHES A SWALLOWTAIL BUTTERFLY, WHICH THE BIRD TYPICALLY SNARES WHILE BOTH ARE IN FLIGHT. BIRDS LEARN TO STEER CLEAR OF SOME POTENTIAL PREY LIKE THIS TUSSOCK MOTH CATERPILLAR *(RIGHT)*, WHOSE STIFF, BRISTLY HAIRS CAN CAUSE EXTREME IRRITATION WHEN TOUCHED.

THE GLOW OF DAWN MAKES A MANED WOLF IN BRAZIL, A JACKRABBIT IN CALIFORNIA, AND AN ORB SPIDER'S WEB IN WISCONSIN ALL APPEAR AS IF DUSTED WITH GOLD. SOME ORB WEAVER SPIDERS SPIN A NEW WEB EVERY DAY, AFTER FIRST EATING THE PREVIOUS DAY'S WEB.

COLORS

IN SOUTH AFRICA, A CHAMELEON *(LEFT)* STEPS GINGERLY ACROSS A HOT ROAD, PERHAPS UNAWARE OF THE GRASSHOPPER THAT FINDS PIGGYBACKING AN EASIER WAY TO TRAVEL.

A GRASSHOPPER NYMPH *(BELOW)* CAN "HIDE IN PLAIN SIGHT" BY ADJUSTING ITS COLOR TO MATCH ITS SURROUNDINGS.

ON MADAGASCAR, AN ISLAND OFF AFRICA'S EAST COAST, A VERDANT TREE ACCENTS LIMESTONE PINNACLES REACHING HEIGHTS OF UP TO 100 FEET.

PREVIOUS PAGE: THE CHAMELEON CAN CHANGE FROM GREEN TO OTHER COLORS IN RESPONSE TO LIGHT, TEMPERATURE, OR MOOD.

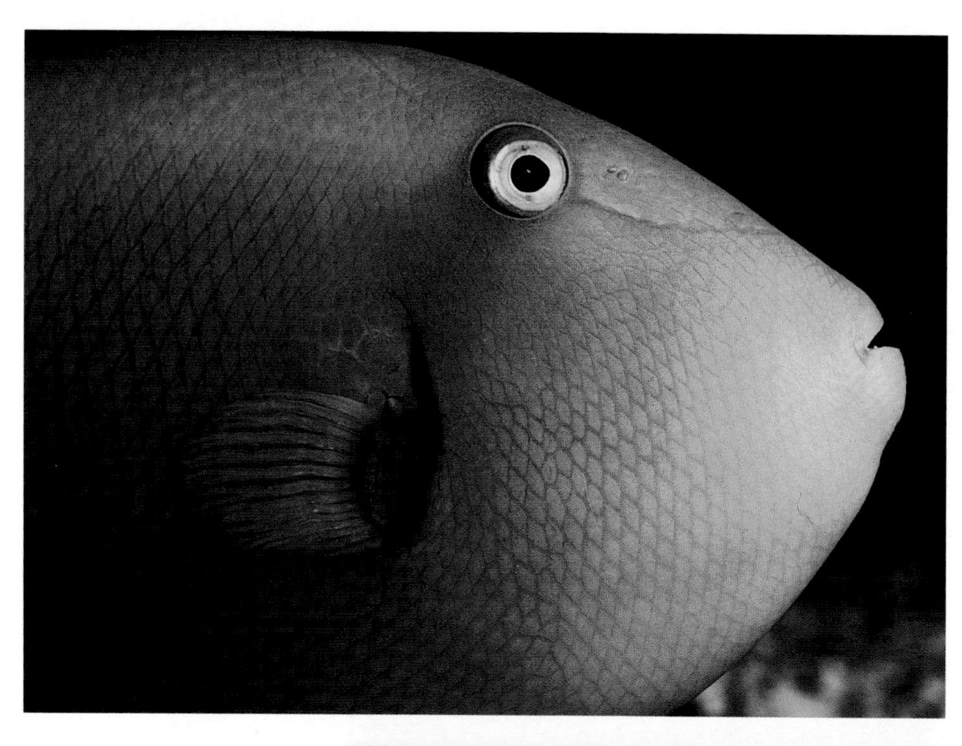

THE PINK-TAILED TRIGGER-FISH *(LEFT)* IS ACTUALLY MOSTLY GREEN. TOUGH SKIN IMBEDDED WITH SCALES GIVES THIS FISH A KIND OF PROTECTIVE ARMOR.

THE LEAFLIKE EXPANSION ON THE HEAD OF A CENTRAL AMERICAN MANTID *(BELOW)* HELPS IT BLEND WITH VEGETATION TO AVOID PREDATORS AND TO AMBUSH PREY.

A PUDDLE FROG IN COSTA RICA NAPS SAFELY IN AN UNROLLING LEAF *(RIGHT)*. WITHIN A DAY, THE LEAF WILL COMPLETELY UNFURL, FORCING THE FROG TO FIND ANOTHER RETREAT.

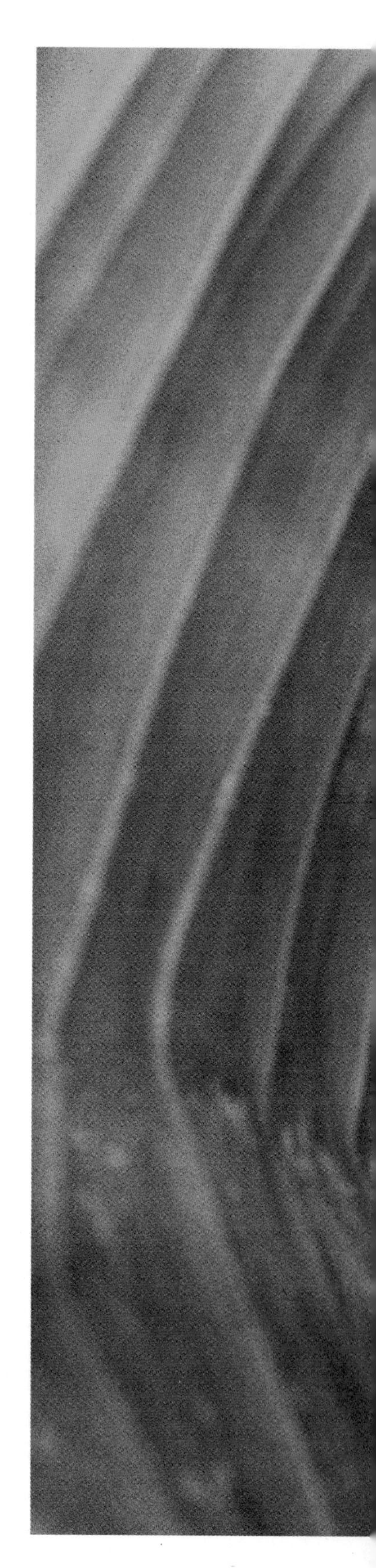

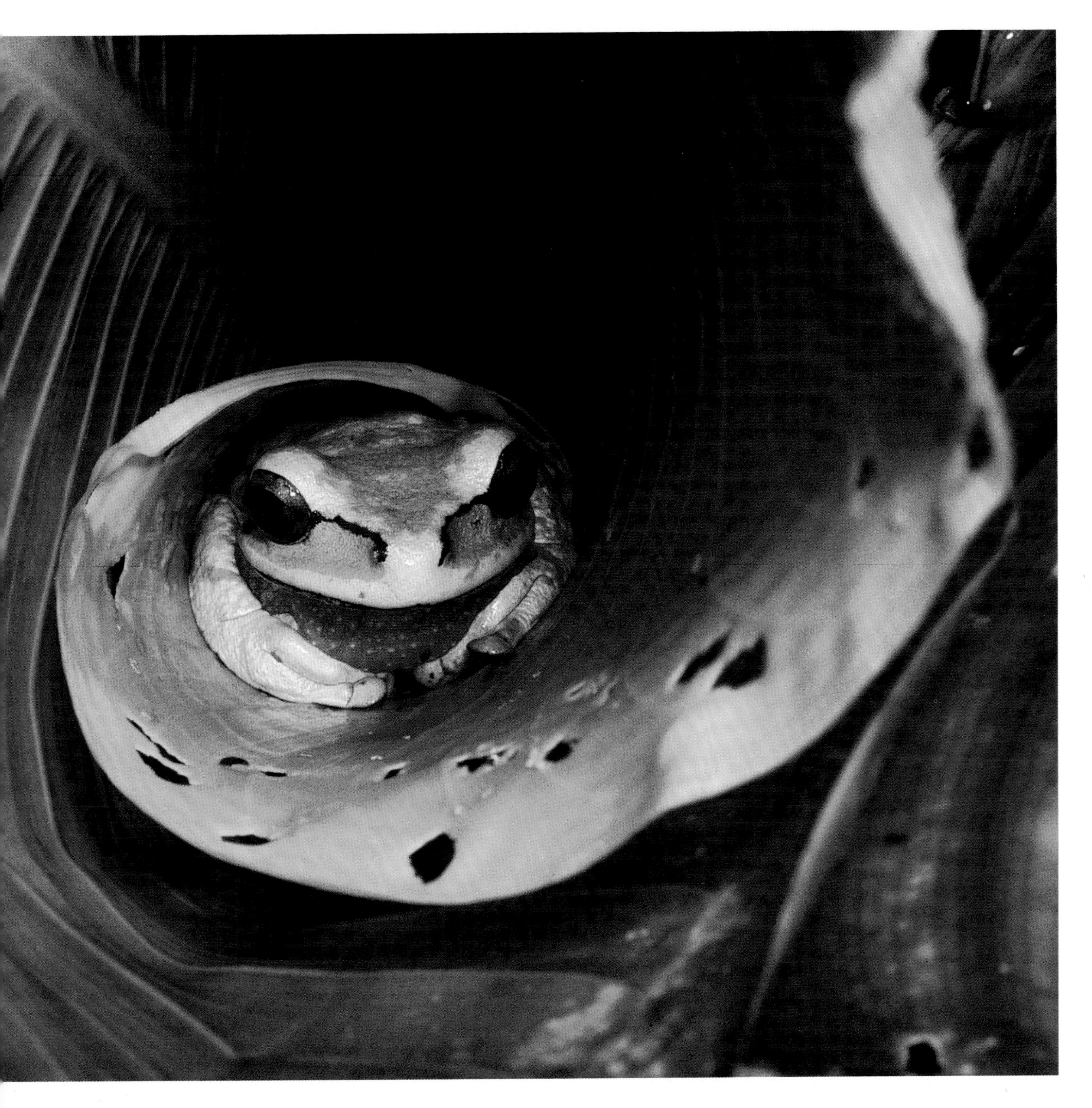

BACKLIGHT ACCENTUATES THE DELICATE FORM AND HUE OF A LUNA MOTH *(RIGHT)*. IN A MANNER TYPICAL OF MOTHS, IT HOLDS ITS WINGS HORIZONTALLY WHEN AT REST. BUTTERFLIES, ON THE OTHER HAND, HOLD THEIR WINGS VERTICALLY WHEN THEY ALIGHT, AS DEMONSTRATED BY A MALACHITE *(LEFT)*.

SUNLIGHT FILTERED THROUGH A CANOPY OF LUSH FOLIAGE IN A CEDAR GROVE CREATES AN INVITING JADE WORLD BELOW. IT IS EASY TO IMAGINE THE SOFT CALLS AND RUSTLINGS OF WARBLERS, THRUSHES, RED SQUIRRELS, AND OTHER RESIDENT ANIMALS.

OVERLAPPING MAYAPPLE LEAVES *(LEFT)* RESEMBLE AN ARRAY OF GREEN UMBRELLAS DRYING AFTER A STORM. THE LEAVES ARE POISONOUS TO HUMANS IF EATEN IN LARGE QUANTITIES.

A CLOSE LOOK AT A SAGUARO CACTUS HIGHLIGHTS ITS SUCCULENT SKIN AND RADIATING SPINES *(RIGHT)*. WINDING ITS WAY AMONG THE SPINES IS A DELICATE, FLOWERING JANUSIA PLANT, WHOSE STEM CAN MEASURE UP TO 10 FEET LONG.

AFTER SURFACING FROM A DIVE, A MUSKRAT SPORTS A HEADDRESS OF DRIPPING DUCKWEED *(LEFT).* MUSKRATS FEAST MAINLY ON PLANTS, BUT DIP DOWN DURING THEIR WANDERINGS TO SNATCH AN OCCASIONAL WATER SNAIL OR MUSSEL.

A MISTY MORNING FINDS A DOE SAFELY ENSCONCED IN THE TALL GRASSES OF A SEDGE MEADOW *(ABOVE).*

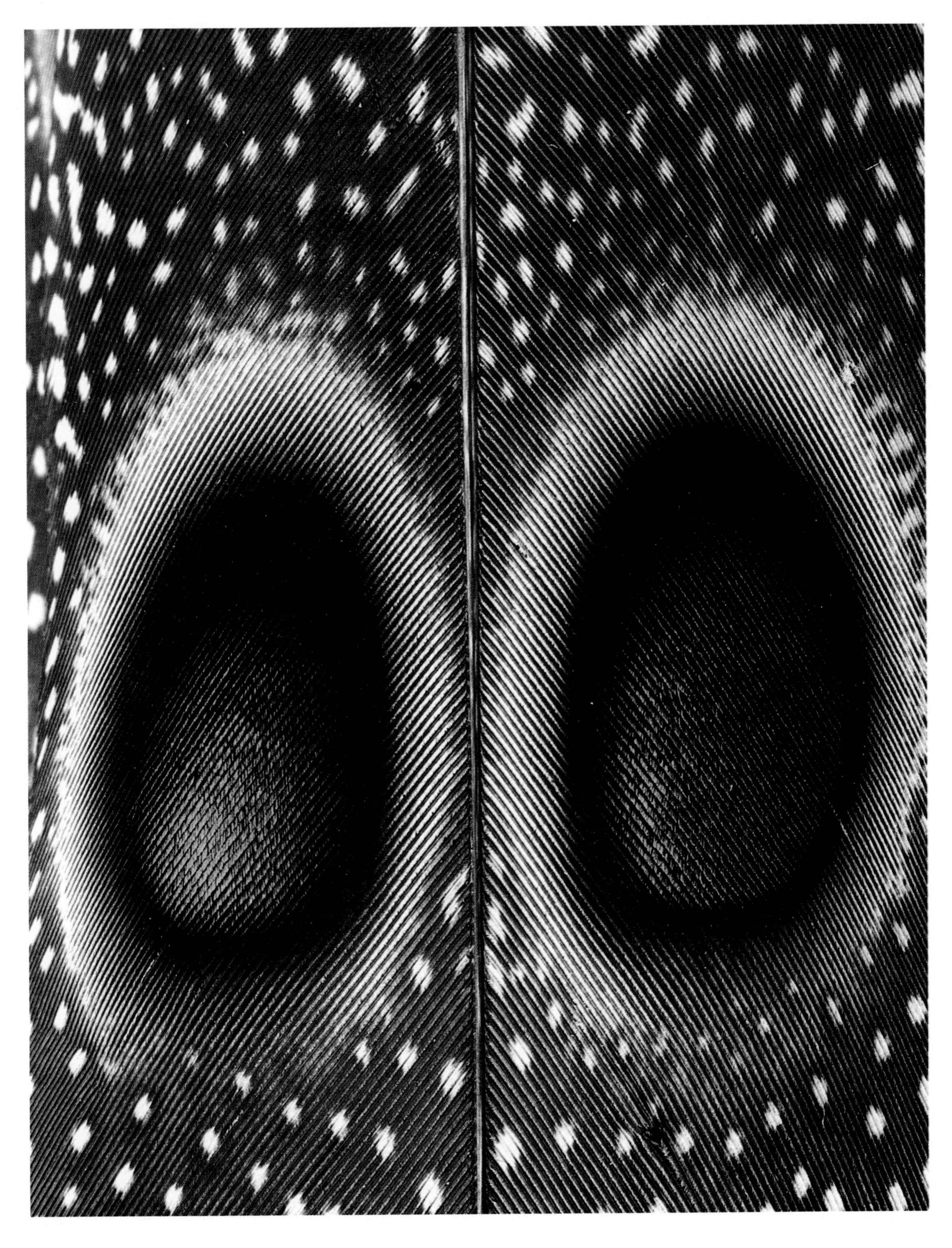

SHIMMERING EMERALD ADORNS THE TAILFEATHERS OF A CENTRAL AMERICAN QUETZAL *(FAR LEFT)* AND THE "EYESPOTS" ON A PEACOCK PHEASANT'S FEATHERS *(LEFT)*. THE IRIDESCENT COLOR IS PRODUCED NOT BY PIGMENT, BUT BY THE WAY LIGHT REFLECTS FROM A THIN COATING OF OIL ON THE FEATHERS.

COLORS

A FEMALE HOODED MERGANSER *(LEFT)* FEATURES THE LONG, ELEGANT HEADFEATHERS FOR WHICH THE DUCK IS NAMED. WHEN THESE FEATHERS ARE RAISED, THE HOOD LOOKS MORE LIKE A CROWN.

BETWEEN FORAYS INTO THE SEA FOR FISH, A COLONY OF FUR SEALS RESTS *EN MASSE* ON THE SHORES OF ALASKA'S PRIBILOF ISLANDS.

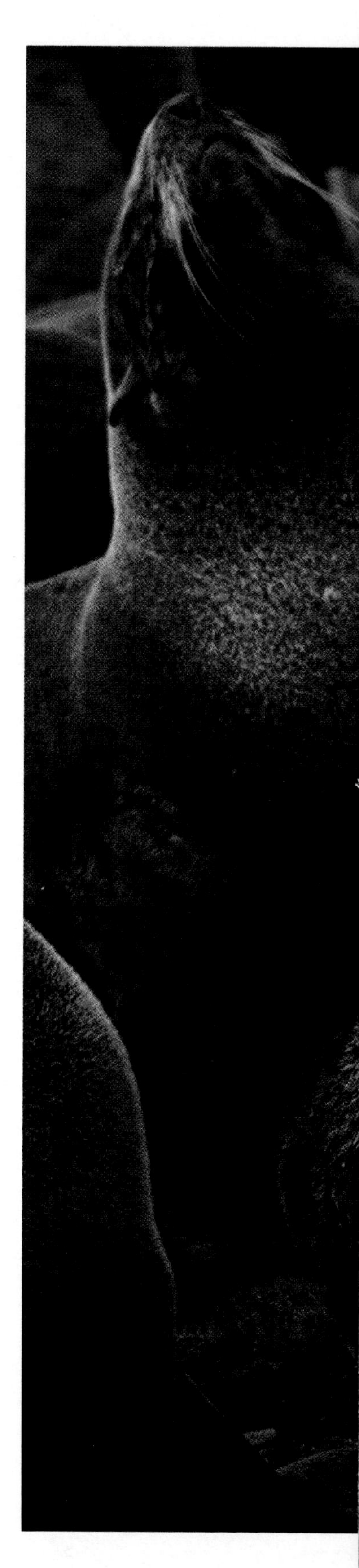

PREVIOUS PAGE: AN ELEPHANT SEAL BASKS IN THE SUN ON BIRD ISLAND, NEAR ANTARCTICA.

SUBSTANTIAL AUTUMN SNOWS PROMPT BROWN BEARS LIKE THESE CUBS *(LEFT)* TO FIND AND ENTER DENS FOR THEIR WINTER SLEEP. THE ONLY CONCESSION ELK IN YELLOWSTONE *(RIGHT)* MAKE TO WINTER IS MIGRATION TO ANOTHER RANGE WHERE THE WEATHER IS ONLY A BIT WARMER AND DRIER.

PRAIRIE DOGS, AMONG THE MOST SOCIAL OF MAMMALS, CAN BECOME BITING, SCRATCHING BALLS OF FURY WHEN MEMBERS OF DIFFERENT SOCIAL GROUPS CHALLENGE EACH OTHER OVER TERRITORY *(LEFT)*.

THOUGH ABLE TO FLY, THE ROADRUNNER *(RIGHT)* PREFERS TO RELY ON ITS GROUND SPEED—UP TO 20 MILES PER HOUR—TO CHASE PREY AND EŞCAPE ENEMIES.

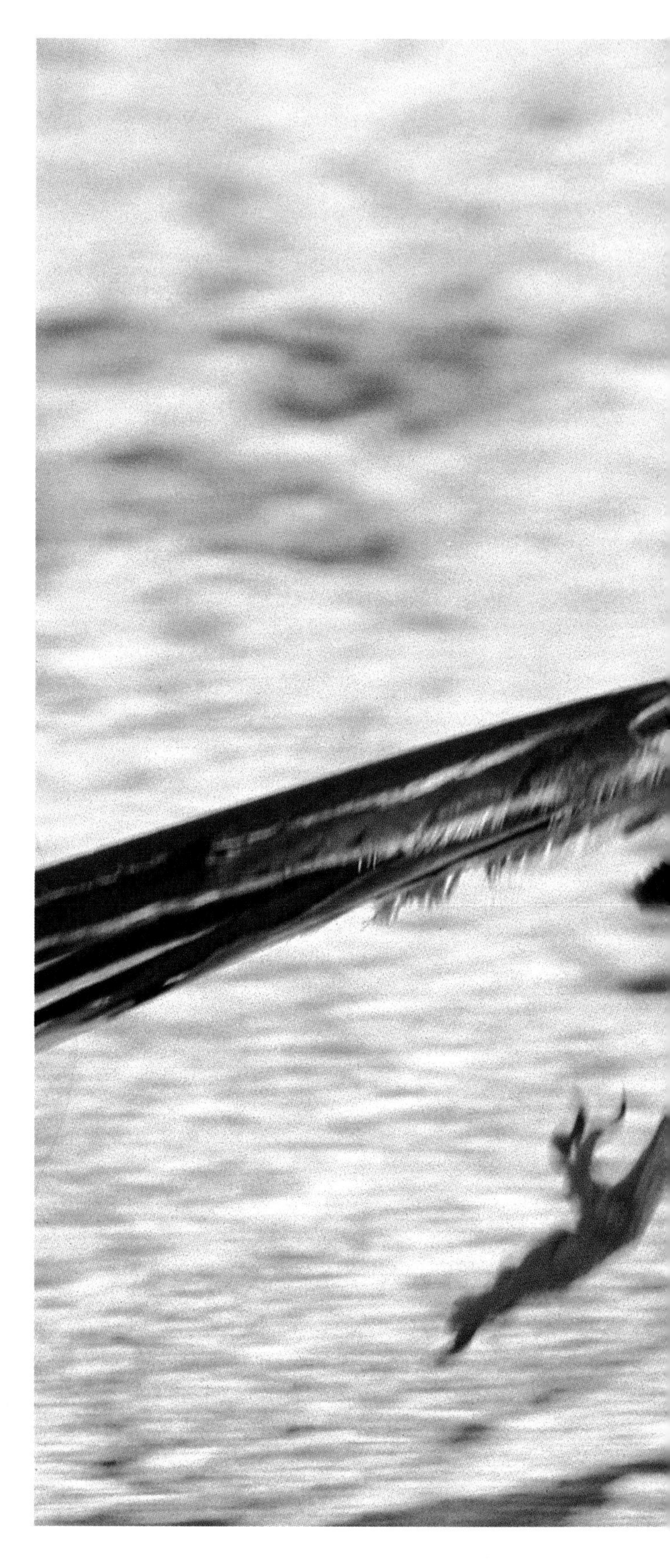

LIKE THE DANDELION, THE GOATSBEARD HAS A SEED-BALL CONTAINING DRY RADIAL PODS INTERLACED WITH SILKY FUZZ.

FORMER HOMES OF SEA SCALLOPS, THESE SHELLS WERE FORMED WHEN LIQUID CALCIUM CARBONATE EXUDED FROM THE SCALLOPS' BODIES AND HARDENED.

THE POLKA-DOTS OF A YOUNG SPOTTED SWEETLIPS *(FAR LEFT)* WILL DISAPPEAR AS THE FISH MATURES. ADULTS ARE GRAYISH BLUE WITH SUBTLE HORIZONTAL LINES.

A CHOCOLATE-BROWN MORAY EEL, HOPING TO SPOT UNWARY PREY, POKES ITS HEAD FROM ITS HIDING PLACE *(LEFT)*. MORAY EELS, INHABITANTS OF CORAL REEFS, CAN ALSO BE YELLOW, GREEN, OR SPOTTED.

A CHAIN OF EGGS RELEASED BY A FEMALE ROCKY MOUNTAIN TOAD FLOATS IN THE WATER NEAR THE MATING COUPLE *(LEFT)*. THE FEMALE, ON THE BOTTOM, IS THE LARGER OF THE TWO.

AMONG BUTTRESSED CYPRESS TRUNKS IN A FLORIDA SWAMP, A WHITE-TAILED BUCK PAUSES ON ALERT FOR SIGNS OF DANGER BEFORE DIPPING HIS HEAD TO DRINK.

A MOOSE COW IN ALASKA NUZZLES ONE OF HER NEWBORN TWINS. THE BIRTHING SPOT NEAR A POND ALLOWS HER TO OBTAIN FOOD AND WATER WHILE REMAINING CLOSE TO HER CALVES.

ASIA'S LEAF FROG *(LEFT)* AND AUSTRALIA'S LEAF MOTH *(FAR LEFT)* SHARE THE ABILITY TO BLEND WITH THEIR EARTHY SURROUNDINGS AS PROTECTION AGAINST PREDATORS. AT HOME IN BEDS OF DEAD LEAVES, THESE ANIMALS ARE STRIKING EXAMPLES OF THE ROLE COLOR AND SHAPE PLAY IN NATURAL MIMICRY.

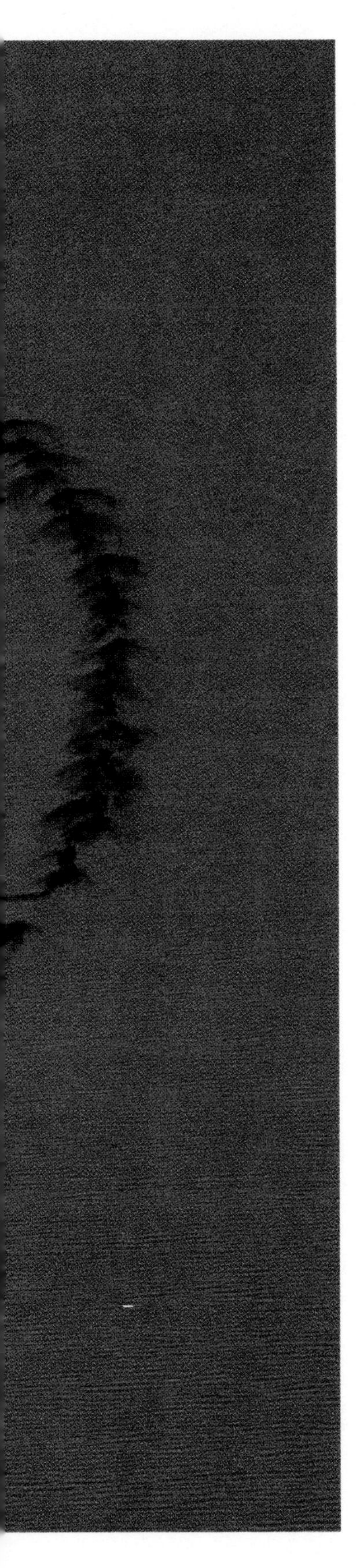

ORYXES LEAVE A SWEEPING TRAIL AS THEY COURSE THE DUNES OF NAMIBIA. IN THE SAME AFRICAN DESERT, A SIDEWINDING ADDER LEAVES AN UNDULATING, LADDER-LIKE TRACK AS IT CROSSES THE SAND. THE "RUNGS" OF THE LADDER ARE NEARLY PERPENDICULAR TO THE DIRECTION OF TRAVEL.

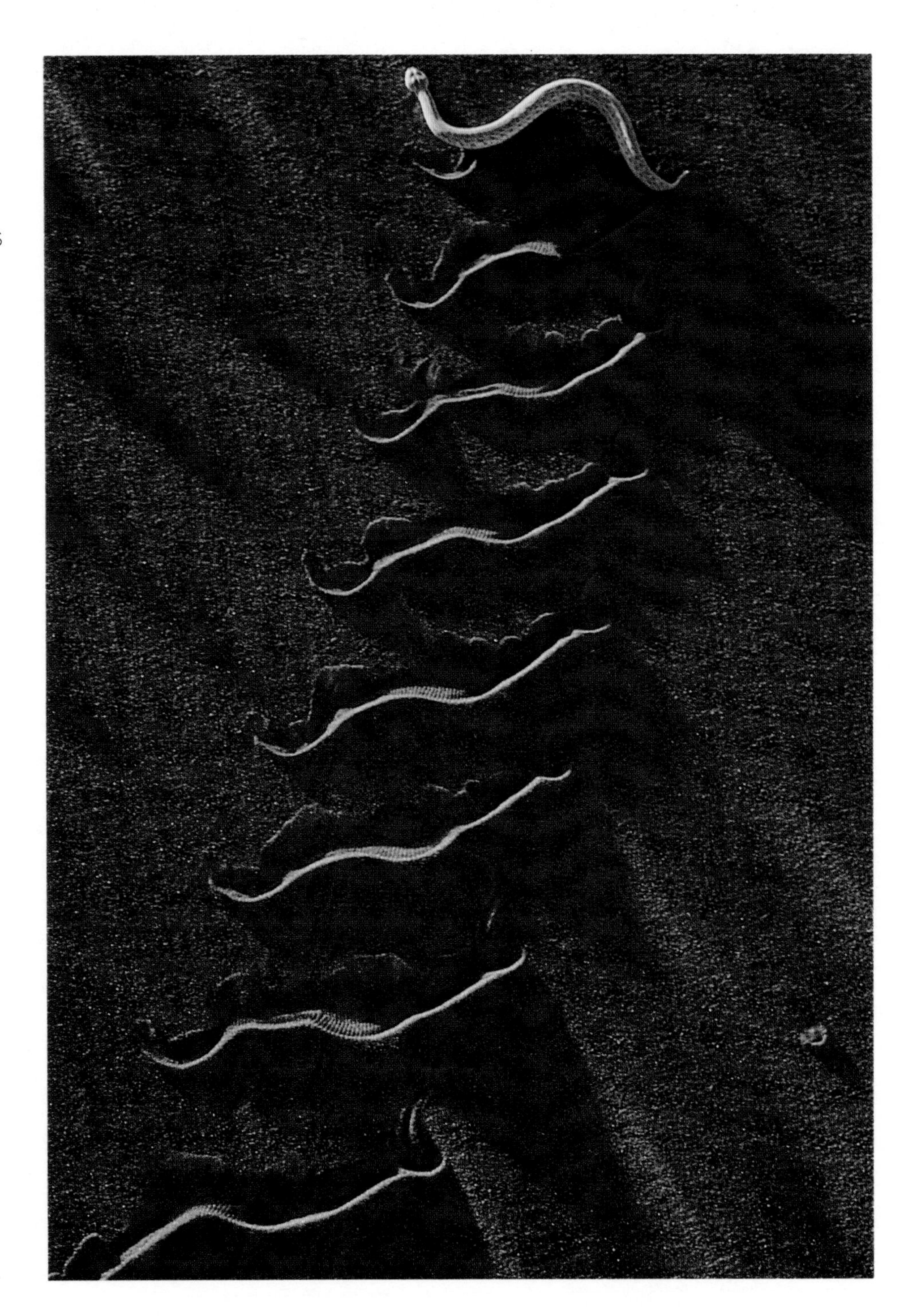

COLORS

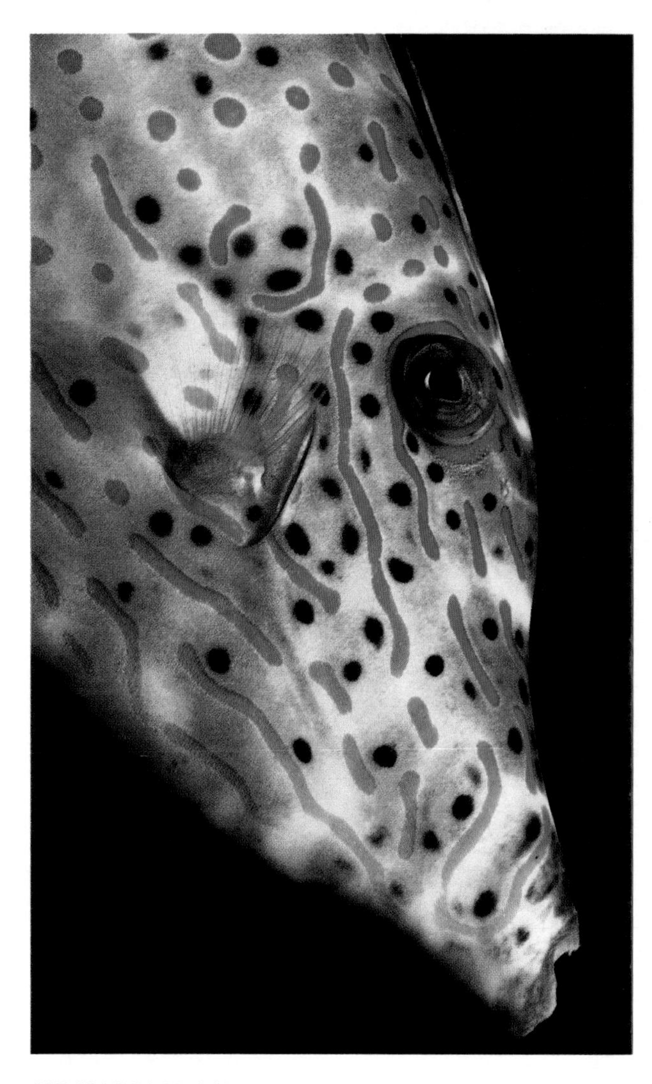

PREVIOUS PAGE: CORAL REEFS TEEM WITH COLORFUL CREATURES LIKE THE ELECTRIC-BLUE DAMSELFISH.

AZURE DOTS AND DASHES ON THE SCRAWLED FILEFISH STAND OUT AGAINST A BACKGROUND SHADE THAT LIGHTENS OR DARKENS ACCORDING TO THE FISH'S SURROUNDINGS AND MOOD.

BIG-EYED JACKS *(RIGHT)*, WEIGHING ABOUT 25 POUNDS EACH, TRAVEL THEIR MARINE WORLD IN SCHOOLS THAT VARY IN SIZE FROM ABOUT A DOZEN TO HUNDREDS OF FISH. THEIR ATTACKS ON SCHOOLS OF SMALLER FISH CAN SET THE WATER CHURNING.

BLUEBIRDS THAT DO NOT MIGRATE MAY HUDDLE TOGETHER FOR WARMTH DURING PARTICULARLY COLD WEATHER *(FAR LEFT).*

THE CLOUDED EYES OF A BARRED OWLET *(ABOVE)* APPEAR BLUE IN BRIGHT LIGHT. AS THE OWLET MATURES, ITS EYES WILL BECOME DARK BROWN.

TALL BILBERRIES *(LEFT),* MEMBERS OF THE BLUEBERRY GROUP, GROW ON MOUNTAIN SLOPES IN THE NORTHERN STATES. THEY MAKE GOOD EATING FOR BLACK BEARS AND HUMANS ALIKE.

In the shallow tropical waters preferred by humpback whales for birthing, a mother and calf stick close together. Scientists have found that humpback mothers and babies are very sensitive to each other, touching and caressing frequently.

A TERN IN FLIGHT IS SILHOUETTED AGAINST THE TAPESTY OF TEXTURES AND REFLECTIONS IN A SOUTH CAROLINA SLOUGH *(ABOVE)*. IN A NORTHWOODS AREA OF WISCONSIN, BIRCH TREES MIRRORED IN A SHADED ICE FLOE CREATE AN ABSTRACT PICTURE IN SAPPHIRE AND GOLD *(RIGHT)*.

SPLAYED PATCHES OF BLUE MARK A COLONY OF BLUE-FOOTED BOOBIES ON THE GALAPAGOS ISLANDS *(FAR LEFT).* DURING MATING SEASON, THE BIRDS DISPLAY THEIR FLASHY FEET IN LEG-LIFTING COURTSHIP DANCES.

AT FIRST, MALE AND FEMALE BOOBIES SHARE PARENTAL DUTIES *(LEFT).* BUT WHEN FEEDING REQUIREMENTS GET HEAVIER, THE FEMALE, WHICH IS LARGER, TAKES OVER.

A FEW SPOTS OF SUNLIGHT ARE THE ONLY WARM TOUCHES IN THIS ICY JANUARY-MORNING SCENE IN YELLOWSTONE NATIONAL PARK'S LAMAR VALLEY.

SOME BLUEJAYS DO NOT MIGRATE SOUTH FOR THE WINTER *(LEFT).* FOOD MAY BE MORE SCARCE THAN IN WARMER MONTHS, BUT SUPPLIES OF BERRIES AND INSECT EGGS—AND PERHAPS ACORNS BURIED BY THE JAY IN THE FALL—ARE USUALLY QUITE SUFFICIENT.

A RISING SUN REVEALS ASPENS ON THE NORTH RIM OF THE GRAND CANYON *(RIGHT)* IN ALTERNATING LAYERS OF BRIGHT LIGHT AND COOL SHADOW.

COLORS

RED-WINGED BLACKBIRDS SWIRL LIKE WINDSWEPT LEAVES AROUND TWO BARE TREES SILHOUETTED BY A LOUISIANA SUNSET *(LEFT).*

THE ANHINGA OR "SNAKE BIRD" OFTEN SWIMS WITH ONLY ITS SINUOUS HEAD AND NECK ABOVE WATER. BECAUSE ITS FEATHERS DO NOT REPEL WATER, THE ANHINGA MUST DRY ITS WINGS IN THE SUN *(RIGHT).*

PREVIOUS PAGE: SPAWNING BLOOD STARS ENGAGE IN A COURTSHIP DANCE.

ITS BRIGHT BANDS OF ORANGE HELP A HARMLESS TROPICAL GLOSSY SNAKE *(ABOVE)* MIMIC THE POISONOUS BI-COLORED CORAL SNAKE, THUS FOOLING POTENTIAL PREDATORS.

LIKE THE AUTUMN LEAF IN WHICH SHE HAS DEPOSITED HER EGG SAC, THIS GRASS SPIDER *(RIGHT)* IS AT THE END OF HER LIFE, BUT SHE WILL CONTINUE TO GUARD HER EGGS UNTIL SHE DIES.

DIMINUTIVE MUSHROOMS CAST TREE-LIKE SHADOWS THROUGH A MAPLE LEAF IN ONTARIO *(FAR RIGHT)*.

EVERY WINTER, MONARCH BUTTERFLIES FROM ALL BUT THE WEST COASTS OF THE UNITED STATES AND CANADA MIGRATE TO SMALL PATCHES OF FOREST IN MEXICO. THE WINTERING MONARCHS SHOWN HERE ORNAMENT AN EVERGREEN NEAR MEXICO CITY.

GRAY AT BIRTH, AMERICAN FLAMINGOES DO NOT ATTAIN THEIR STRIKING CORAL HUE UNTIL THEY ARE ABOUT FOUR YEARS OLD AND READY TO BREED *(LEFT)*.

A WOOD LILY PROVIDES A BEAUTIFUL RESTING PLACE AND A NECTAR SNACK FOR THIS TIGER SWALLOWTAIL BUTTERFLY *(RIGHT)*.

CRIMSON LAVA, REACHING TEMPERATURES OF OVER 2000° FARENHEIT, SPOUTS FROM HAWAII'S MOUNT KILAUEA *(LEFT)*.

LIKE STILL-GLOWING EMBERS, RED ROCK CRABS CLING TO VOLCANIC ROCK ON ESPAÑOLA ISLAND IN THE GALAPAGOS *(BELOW)*.

MAPLE TREES IN FALL FOLIAGE PUNCTUATE A STAND OF WHITE BIRCH IN MICHIGAN *(BELOW)* AND MINGLE WITH EVERGREENS IN A VERMONT FOREST *(RIGHT)*.

THE TUFTED PUFFIN *(LEFT)* GROWS A THICK, COLORFUL SHEATH OVER ITS BEAK AND A WHITE TUFT OF FEATHERS ON ITS HEAD EACH SPRING FOR THE MATING SEASON.

THE RARE GOLDEN TOAD *(RIGHT)* IS NAMED FOR THE MALE OF THE SPECIES. FEMALES MAY BE EITHER CREAM COLORED, OR DARK GREEN OR BLACK WITH RED SPOTS.

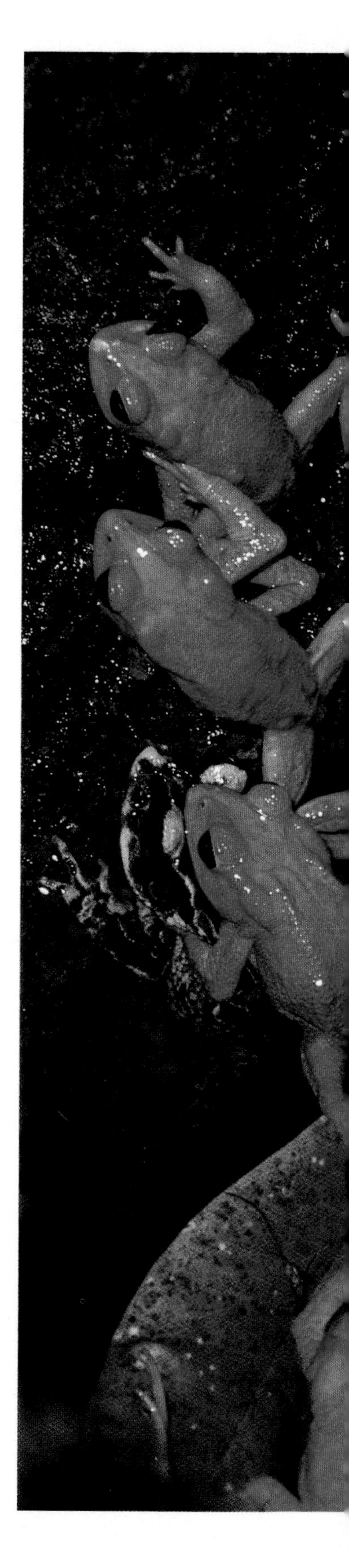

ALTHOUGH IT IS MOST OFTEN SEEN IN THE FAMILIAR ORANGE-RED VARIETY, THE RED FOX *(LEFT)* CAN ALSO BE A RARE SILVER-BLACK OR REDDISH BROWN WITH A BLACK STRIPE ACROSS ITS BACK AND SHOULDERS.

DAWN'S ORANGE GLOW FILTERS THROUGH CYPRESS TREES IN THE ATCHAFALAYA SWAMP, SIGNALING THE BEGINNING OF ANOTHER SUBTROPICAL DAY *(RIGHT)*.

COLORS

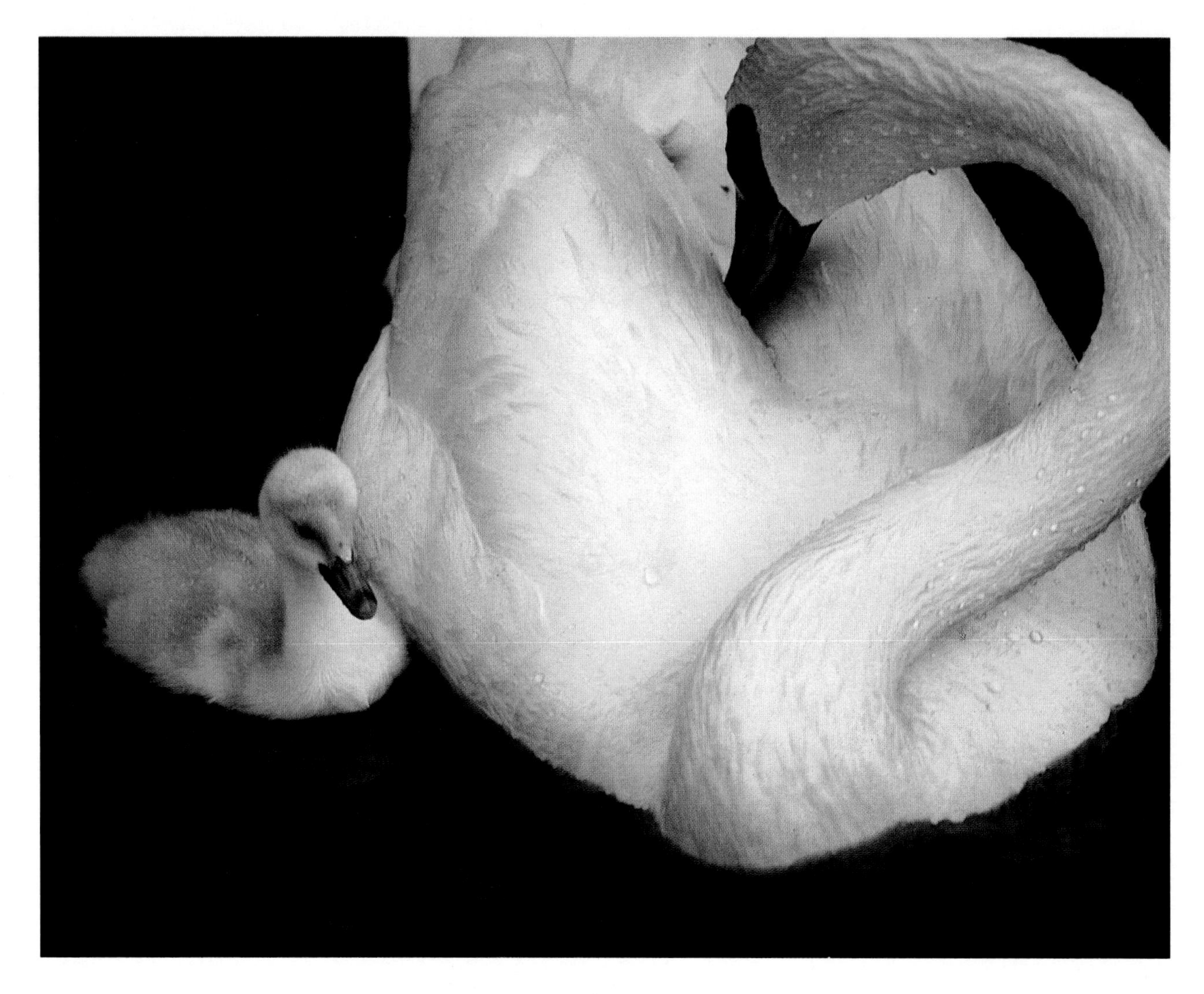

TRUMPETER SWANS *(LEFT)* WEIGH ONLY SEVEN OUNCES AT BIRTH, BUT ADULTS CAN WEIGH UP TO 38 POUNDS AND HAVE A WINGSPAN OF SIX TO EIGHT FEET, MAKING THEM THE LARGEST SWANS IN THE WORLD.

THE GREAT EGRET *(RIGHT)* IS NAMED FOR ITS GRACEFUL PLUMES, OR "AIGRETTES," WHICH BEGIN TO GROW EACH YEAR JUST BEFORE THE MATING SEASON.

PREVIOUS PAGE: TRUMPETER SWAN CYGNETS PADDLE CONTENTEDLY IN GRAND TETON NATIONAL PARK.

THE DASHING DOWNY WOODPECKER *(LEFT)*, A MERE SIX INCHES LONG, IS THE SMALLEST AND MOST COMMON OF ALL NORTH AMERICAN WOODPECKERS.

THEORIES ON THE PURPOSE OF THE ZEBRA'S STRIPED COAT *(RIGHT)* DIFFER. IT MAY SERVE AS A RECOGNITION SIGNAL FOR OTHER ZEBRAS, AS CAMOUFLAGE, OR AS A FLY DETERRENT.

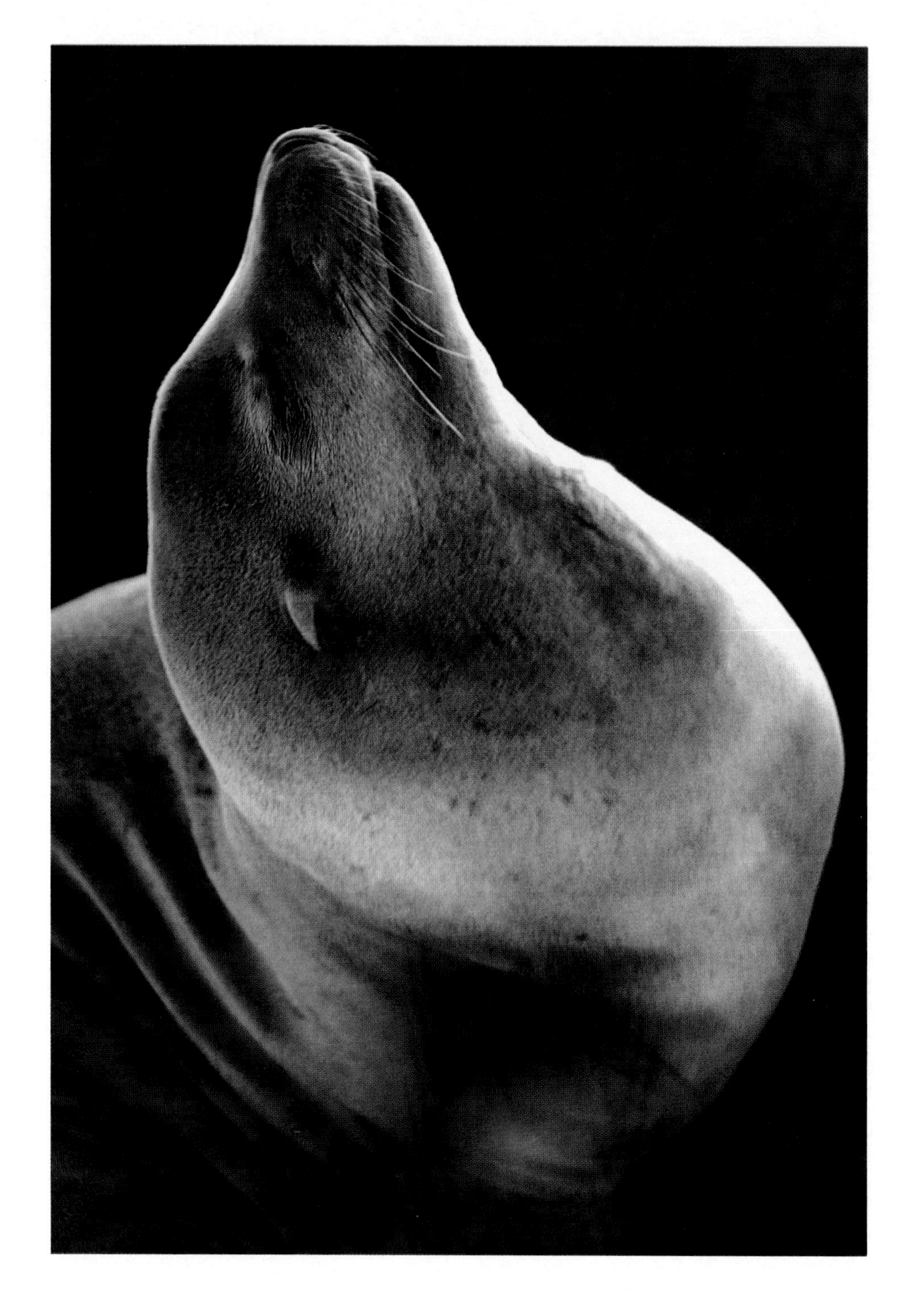

A CALIFORNIA SEA LION ON THE MONTEREY PENINSULA BASKS SENSUOUSLY IN THE SUN *(LEFT).* LIKE ALL PINNIPEDS, IT HAS SPECIALIZED VERTEBRAE WHICH MAKE ITS THICK, MUSCULAR NECK SURPRISINGLY FLEXIBLE.

FROST GIVES THIS ENGLISH IVY A CONFECTIONARY APPEARANCE *(RIGHT).* FOLKLORE HAS IT THAT IVY CAN REMEDY HANGOVERS AND WARD OFF WITCHES.

SHROUDED IN MORNING MIST, LIGHT GRAY SANDHILL CRANES LOOK BLACK. SANDHILLS OFTEN APPEAR RUST COLORED BECAUSE IRON IN TUNDRA PONDS STAINS THEIR FEATHERS.

A LICHEN-COVERED OAK SHINES WITH AN EERIE BEAUTY UNDER A COLD DECEMBER MOON *(LEFT).*

THIS FROST-COVERED QUEEN ANNE'S LACE *(RIGHT)* HAS CONTRACTED INWARD WHILE ITS SEEDS MATURE. THE FLOWER'S APPEARANCE AT THIS STAGE HAS GIVEN IT THE COMMON NAME OF "BIRD'S NEST."

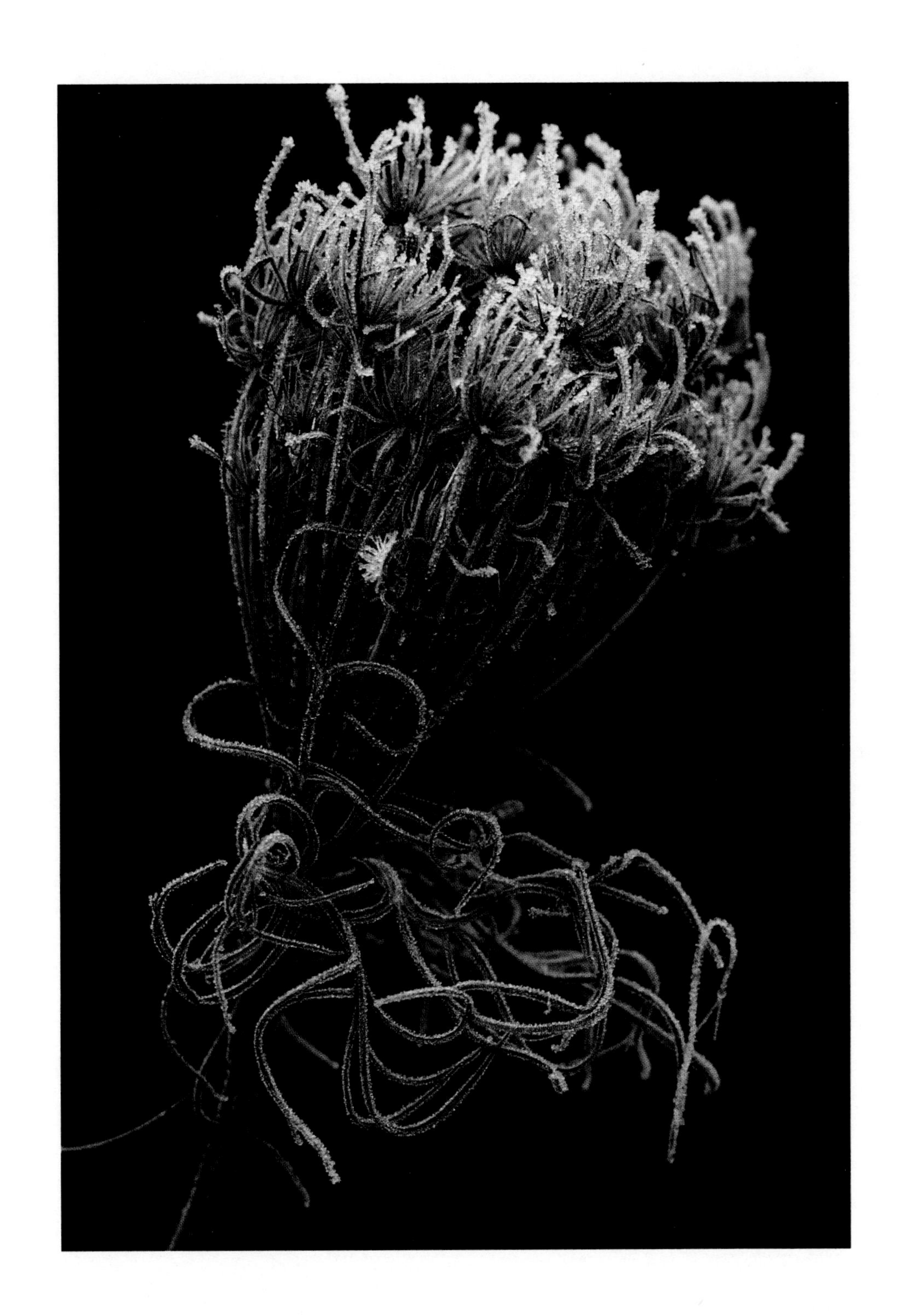

ONCE THEY LEAVE THEIR NESTS, THESE BLACK-BROWED ALBATROSS CHICKS *(ABOVE)* WILL LIVE ON THE HIGH SEAS, RETURNING TO LAND ONLY WHEN IT IS TIME FOR THEM TO BREED.

SOME PENGUIN COLONIES HAVE MORE THAN A MILLION MEMBERS. HERE, A SMALL GROUP OF ADÉLIE PENGUINS LINES UP FOR A BRISK SWIM IN ANTARCTIC WATERS *(RIGHT)*.

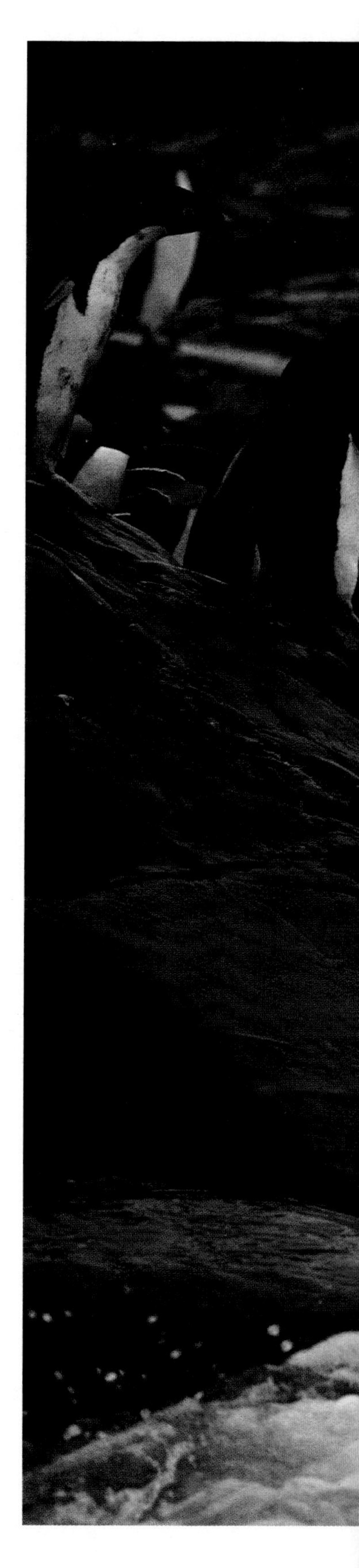

DESPITE THEIR REPUTATION, SKUNKS SPRAY ONLY WHEN PROVOKED. THE HOODED SKUNK *(LEFT)* IS PARTICULARLY SECRETIVE. ITS NOCTURNAL HABITS HELP IT AVOID CONTACT WITH HUMANS.

THE SURVIVAL OF THE ENDANGERED GIANT PANDA *(RIGHT)* LARGELY DEPENDS ON ITS ABILITY TO OBTAIN SUFFICIENT QUANTITIES OF BAMBOO, ITS MAIN FOOD.

COLORS

NUDIBRANCHS, OR SEA SLUGS *(LEFT)*, ARE COLORFUL COUSINS OF LAND SLUGS. THEIR BRIGHT HUES WARN PREDATORS OF THE POISONOUS, STINGING "HORNS" THAT PROTRUDE FROM THE NUDIBRANCH'S BODY.

CLOWNFISH SPEND MOST OF THEIR TIME PROTECTED AMONG THE POISONOUS BRANCHES OF SEA ANEMONES *(RIGHT)*, WHICH OTHER FISH AVOID. A THICK, SLIMY COATING ON CLOWNFISH SOMEHOW RENDERS THEM IMMUNE TO THE ANEMONE'S STING.

PREVIOUS PAGE: DEEP PINK MAY-BLOOMING FLOWERS ENCIRCLE THE BARREL OF A FISHHOOK CACTUS IN ARIZONA.

TINY AND DELICATE, THESE MARASMIUS MUSHROOMS *(LEFT)* HAVE CAPS ONLY ABOUT ONE-HALF INCH ACROSS. THEY ARE COMMON IN LOWLAND RAIN FORESTS OF CENTRAL AND SOUTH AMERICA AND GROW ONLY ON ROTTING LEAVES.

A BABY SNOW MONKEY FINDS SOME RESPITE FROM THE COLD BY SNUGGLING AGAINST ITS MOTHER'S THICK FUR *(RIGHT)*. SNOW MONKEYS, OR JAPANESE MACAQUES, LIVE IN THE FORESTED MOUNTAINS OF NORTHERN JAPAN WHERE WINTER TEMPERATURES OFTEN DIP BELOW 20° F.

FRESHLY MOLTED EASTERN WHITE PELICANS IN INDIA *(LEFT)* PREEN THEIR NEW PLUMAGE, WHICH IS TINGED A DELICATE ROSY PINK.

STARFISH KNOWN AS OCHRE SEA STARS CAN BE VARIOUS SHADES OF PURPLE AS WELL AS ORANGE AND YELLOW *(RIGHT)*. COMMON ALONG THE PACIFIC COAST, THEY ARE ESPECIALLY NOTICEABLE ON ROCKY SHORES WHEN THE TIDE IS OUT.

THE REFLECTION OF A LAVENDER AND PINK TWILIGHT SKY OVER NORTH CAROLINA'S LAKE MATTAMUSKEET IS ETCHED WITH THE SILHOUETTES OF AQUATIC PLANTS AND CYPRESS TREES.

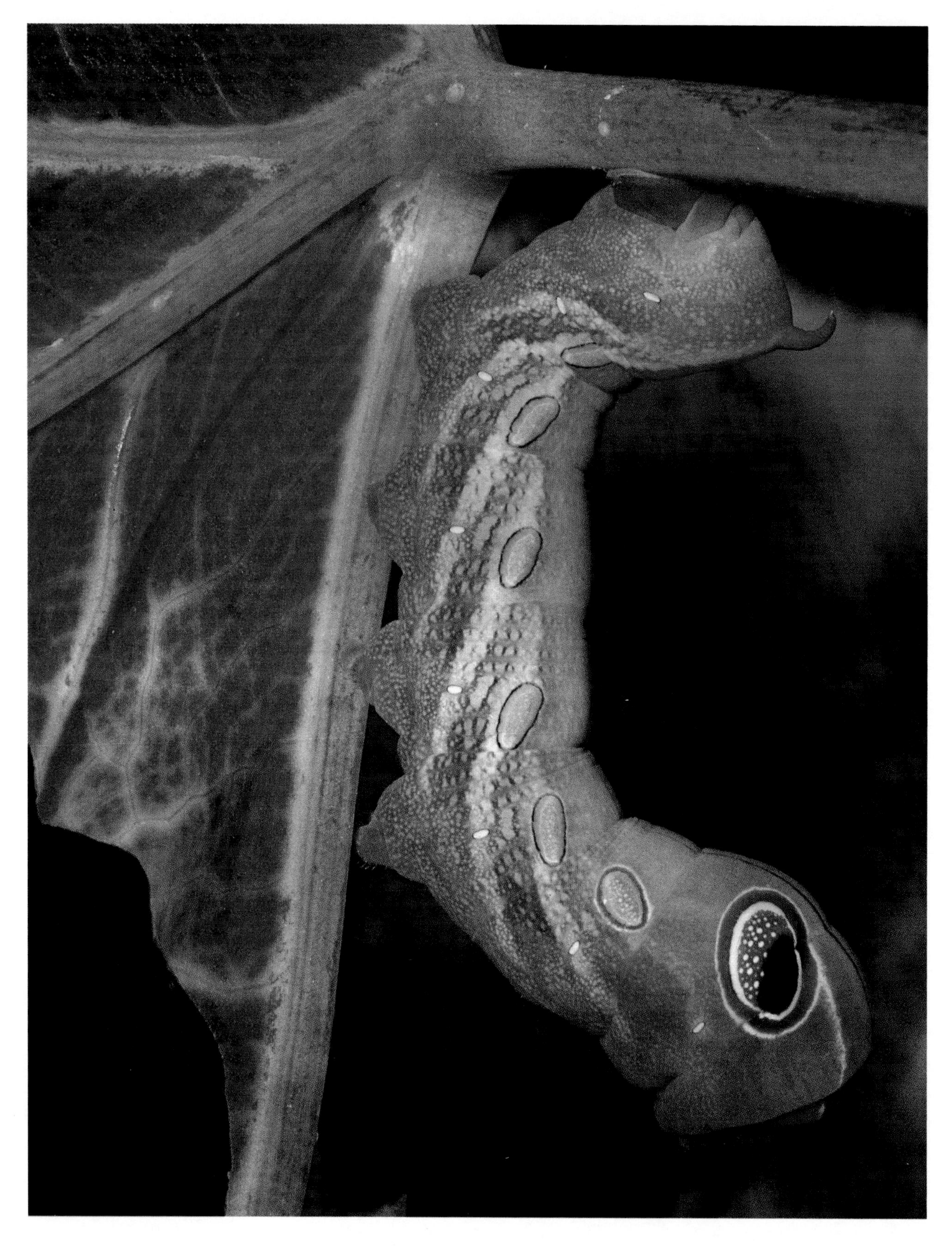

HAWK MOTH CATERPILLARS *(LEFT)* ARE MARKED WITH DARK "EYESPOTS" THAT INTIMIDATE PREDATORS, WHO PROBABLY MISTAKE THE SPOTS FOR THE EYES OF A DIFFERENT AND MORE FORMIDABLE CREATURE.

THE HAWK MOTH'S LARGE BODY *(RIGHT)* AND ITS HABIT OF HOVERING IN FRONT OF FLOWERS WHILE FEEDING ON NECTAR LEAD SOME PEOPLE TO MISTAKE IT FOR A HUMMINGBIRD.

COLORS

IN AUSTRALIA, A RAINBOW LORIKEET CRUISES THE BLOSSOMING WATTLE TREES IN A QUEENSLAND FOREST *(LEFT)*. THE FOOT-LONG BIRD FEEDS ON THE BLOSSOMS' NECTAR, WHICH IT EASILY EXTRACTS WITH THE BRUSHLIKE TIP OF ITS TONGUE.

RED POPPIES, ORANGE WALLFLOWERS, PURPLE CORNFLOWERS, AND WHITE BABY'S BREATH COVER A SPRING MEADOW WITH A PROFUSION OF COLOR.

PREVIOUS PAGE: A RED-EYED TREE FROG'S SUDDEN SHOW OF COLORS CAN STARTLE ATTACKERS, GIVING THE TINY CREATURE A CHANCE TO ESCAPE.

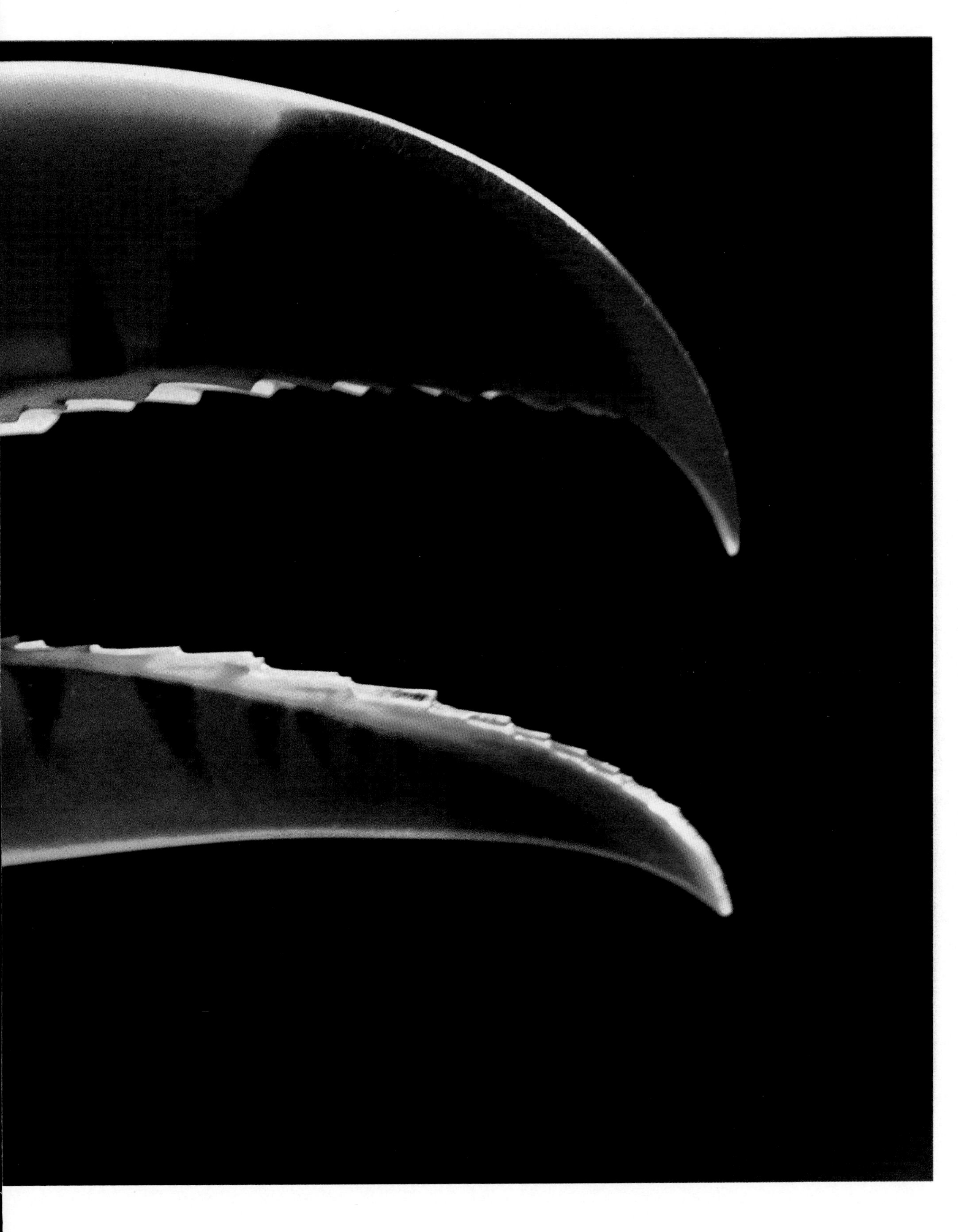

THE KEEL-BILLED TOUCAN, WITH ITS MULTI-HUED BILL, IS ONE OF THE MOST COLORFUL OF THE TOUCAN FAMILY. ALMOST AS LONG AS ITS BODY, THE TOUCAN'S BILL CONSISTS OF A THIN, HARD SHEATH COVERING HONEYCOMBED BONE, MAKING IT STRONG BUT LIGHT.

LINING TROPICAL COASTS ARE COLORFUL CORAL REEFS THAT ABOUND WITH FISH AS VIBRANT AS THEIR ENVIRONMENT. A PARROTFISH *(LEFT)*, WITH ITS BEAK-LIKE MOUTH, NIBBLES AT PLANTS GROWING ON THE CORAL'S SURFACE, AND ORANGE FAIRY BASSLETS MANEUVER AROUND SOFT CORAL IN THE RED SEA *(RIGHT)*.

THE WING OF A URANIA MOTH IN MADAGASCAR *(LEFT)* AND A GATHERING OF HARLEQUIN BUGS IN AUSTRALIA *(RIGHT)* DAZZLE THE EYE WITH RADIANT COLOR CALLED IRIDESCENCE. IT IS CAUSED BY THE REFLECTION OF LIGHT'S MANY COLORS FROM MULTIPLE SURFACES SUCH AS THE LAYERS IN THE MOTH'S WING SCALES OR THE FACETS OF THE BUGS' UNEVENLY TEXTURED COVERINGS.

A RESPLENDENT MOSAIC OF INDIAN PAINTBRUSH AND TEXAS BLUEBELLS COVERS THE PRAIRIE BELOW A STAND OF LIVE OAKS IN TEXAS.

FOLLOWING PAGE: IN TYPICAL SLEEP POSTURE, A SCARLET MACAW, WITH ITS HEAD TURNED BACKWARDS, TUCKS ITS BEAK UNDER A FOLDED WING. THE BIRD'S OPEN EYE INDICATES IT IS IN A STATE OF WARY SEMI-SLEEP.

CREDITS

Cover: Robert C. Simpson/ Tom Stack and Associates. **Cover flap:** Michael and Patricia Fogden. **Page 1:** Günter Ziesler. **2-3:** Robert C. Simpson/Tom Stack and Associates. **10-11:** Rod Planck/Tom Stack and Associates. **12-13:** Art Wolfe. **13:** Dwight R. Kuhn. **14:** Mary Clay/Tom Stack and Associates. **15:** John Shaw. **16:** Chris Newbert. **17:** Don and Pat Valenti/ DRK Photo. **18 top:** Edward Ross; **bottom:** Michael Fogden/Bruce Coleman, Ltd. **18-19:** Kjell Sandved/ Sandved and Coleman, D.C. **20:** Johnny Johnson. **21:** Tim Fitzharris. **22-23:** Robert P. Carr. **23:** Larry West. **24:** Christian Zuber/Bruce Coleman, Ltd. **25:** Rita Summers. **26-27:** Hans Pfletschinger/Peter Arnold, Inc. **28:** Rod Planck/Tom Stack and Associates. **28-29:** Larry West. **30:** John Shaw. **31 left:** Rod Planck; **right:** Jeff Foott. **32:** David Cavagnaro/DRK Photo. **33:** Rod Planck. **34:** Art Wolfe. **35 left:** Jonathan Scott/Planet Earth Pictures; **right:** Rod Planck. **36:** Jeff Rotman. **37:** Chris Newbert. **38:** Rod Planck. **39:** Fritz Prenzel/Bruce Coleman, Ltd. **40-41:** Larry West. **42:** Leonard Zorn. **43:** Dieter and Mary Plage/Bruce Coleman, Ltd. **44:** Richard Matthews/Planet Earth Pictures. **45 top:** David Cavagnaro; **bottom:** Stephen J. Krasemann/DRK Photo. **46-47:** Dwight R. Kuhn. **48 top:** Anthony Bannister Photo Library; **bottom:** Edward Ross. **49:** Frans Lanting. **50 top:** Frans Lanting; **bottom:** Philip K. Sharpe/Oxford Scientific Films. **50-51:** Michael Fogden. **52:** Gary Braasch. **53:** Stan Osolinski. **54-55:** Pat O'Hara/DRK Photo. **56:** Robert P. Carr. **57:** Willard Clay. **58-59:** Dr. E.R. Degginger. **59:** John Shaw. **60:** David Cavagnaro/DRK Photo. **61:** Frans Lanting. **62-63:** Jim Mastro. **64:** Tim Fitzharris. **64-65:** Bob Krist. **66:** Johnny Johnson. **67:** Daniel J. Cox. **68:** Jerry Ferrara. **68-69:** Phil and Loretta Hermann. **70:** Geoff du Feu/Planet Earth Pictures. **71:** Antony Joyce/Planet Earth Pictures. **72:** A. Kerstitch/Planet Earth Pictures. **73 top:** C.C. Lockwood; **bottom:** Entheos. **74-75:** Erwin and Peggy Bauer/Bruce Coleman, Inc. **75:** Johnny Johnson. **76:** David P. Maitland/ Planet Earth Pictures. **77:** Dr. E.R. Degginger. **78-79:** Jim Brandenburg. **79:** Anthony Bannister Photo Library. **80-81:** Jane Burton/Bruce Coleman, Ltd. **82:** Chris Newbert. **82-83:** Chris Newbert. **84:** Michael Smith. **85 top:** Art Wolfe; **bottom:** Larry West. **86-87:** Deborah Glockner-Ferrari. **88:** Jim Brandenburg. **89:** R. Hamilton Smith. **90:** Frans Lanting. **91:** Tui de Roy. **92-93:** Stan Osolinski. **94:** Larry West. **95:** Pat O'Hara/DRK Photo. **96-97:** Howard Hall. **98:** C.C. Lockwood/DRK Photo. **99:** Stan Osolinski. **100 left:** Michael Fogden; **right:** Robert Noonan. **101:** David Cavagnaro. **102-103:** Frans Lanting. **104:** Richard Kolar/Animals Animals. **105:** Rod Planck. **106-107:** Paul Chesley/ Photographers Aspen. **107:** Günter Ziesler. **108:** Robert P. Carr. **109:** Willard Clay. **110:** M.A. Chappell/ Animals Animals. **110-111:** Michael Fogden. **112:** Peter Bengeyfield. **113:** C.C. Lockwood. **114-115:** Jeff Foott. **116:** Jeff Foott. **117:** Wendel Metzen. **118:** Larry West. **118-119:** Reinhard Künkel. **120:** Robert Noonan. **121:** David Phillips/Planet Earth Pictures. **122-123:** Robert P. Carr. **124:** Peter Menzel. **125:** Skip Moody. **126:** Frans Lanting. **126-127:** M.P. Kahl/Bruce Coleman, Ltd. **128:** Marty Cordano/DRK Photo. **129:** R.Y. Kaufman/Yogi. **130-131:** T.A. Wiewandt. **132:** Howard Hall. **133:** Carl Roessler/Bruce Coleman, Ltd. **134:** Michael and Patricia Fogden. **135:** Dominique Braud. **136:** Fred Bavendam/Peter Arnold, Inc. **137:** Entheos. **138-139:** John Shaw. **140:** Michael Fogden. **141:** Bill L. Ivy. **142-143:** Michael Fogden. **144:** John Cancalosi/Tom Stack and Associates. **145:** Frank Oberle. **146-147:** Michael and Patricia Fogden. **148:** Douglas Faulkner/Sally Faulkner Collection. **149:** Chris Newbert. **150:** Stephen Dalton/Oxford Scientific Films. **150-151:** David P. Maitland/Planet Earth Pictures. **152-153:** John Shaw. **154:** Steve Smith. **Deluxe binding endpapers:** John Gerlach/DRK Photo

Library of Congress CIP Data:

Main Entry Under Title:
Colors in the Wild.
p. cm.
ISBN 0-912186-95-X (trade)
ISBN 0-912186-98-4 (deluxe)
1. Color of animals.
2. Plants—Color. I. National Wildlife Federation.
QL767.C73 1988
591.5—dc 19 88-15162 CIP

STAFF FOR THIS BOOK

Howard F. Robinson
Editorial Director

Bonnie Stutski
Photo Editor and Project Coordinator

Donna Miller
Design Director

Elaine S. Furlow
Bonnie Stutski
Michele Morris
Writers

Michele Morris
Editorial Assistant

Paul Wirth
Quality Control

Margaret E. Wolf
Permissions Editor

Kathleen Furey
Tina Isom
Production Artists

Vi Kirksey
Christine Irwin
Production Assistants

SCIENTIFIC CONSULTANTS

Dr. John A. Allen
George Mason University

Dr. Todd Cook
University of Maryland

Dr. Donald Messersmith
University of Maryland

Dr. S. Douglas Miller
National Wildlife Federation